Frédéric Hecht • Gontran Lance • Emmanuel Trélat

PDE-constrained optimization within FreeFEM

Frédéric Hecht
Laboratoire Jacques-Louis Lions
Sorbonne Université
Paris Cedex 05, France

Gontran Lance
Rond-Point de l'échangeur de Solaize
IFP Energies Nouvelles
Solaize, France

Emmanuel Trélat
Laboratoire Jacques-Louis Lions
Sorbonne Université
PARIS CEDEX 05, Paris, France

ISSN 2731-7595 ISSN 2731-7609 (electronic)
SpringerBriefs on PDEs and Data Science
ISBN 978-981-95-7048-5 ISBN 978-981-95-7049-2 (eBook)
https://doi.org/10.1007/978-981-95-7049-2

Mathematics Subject Classification: 49M41

This Springer imprint is published by the registered company Springer Nature Singapore Pte Ltd.
The registered company address is: 152 Beach Road, #21-01/04 Gateway East, Singapore 189721, Singapore

SpringerBriefs on PDEs and Data Science

SpringerBriefs on PDEs and Data Science targets contributions that will impact the understanding of partial differential equations (PDEs), and the emerging research of the mathematical treatment of data science.

* * *

The series will accept high-quality original research and survey manuscripts covering a broad range of topics including analytical methods for PDEs, numerical and algorithmic developments, control, optimization, calculus of variations, optimal design, data driven modelling, and machine learning. Submissions addressing relevant contemporary applications such as industrial processes, signal and image processing, mathematical biology, materials science, and computer vision will also be considered.

The series is the continuation of a former editorial cooperation with BCAM, which resulted in the publication of 28 titles as listed here: https://www.springer.com/gp/mathematics/bcam-springerbriefs

Preface

This book is intended for students and researchers wishing to learn how to solve constrained optimization problems involving partial differential equations (PDEs) efficiently using the `FreeFEM` software. PDE-constrained optimization problems are frequently encountered in many academic and industrial contexts. Readers should have a basic understanding of the analysis and numerical solving of PDEs using finite element methods, as well as knowledge of optimization algorithms and numerical implementation.

Throughout the book, we examine a variety of examples, some of which are well known and others less so. We have selected these examples to illustrate a variety of situations and hope they will provide a reusable basis for solving other problems in different fields. For each problem, we demonstrate how to effectively use `FreeFEM` and how it can be combined with expert optimization routines such as `IpOpt`, or with automatic differentiation. We provide implementation details and useful tips. All our codes are available on the `FreeFEM` website

https://freefem.org/Optim/

and can, we hope, serve as models for users.

Figure 1, reproduced later in the book, summarizes the various approaches to solving a given PDE-constrained optimization problem. These range from "fully" direct methods (left) to "fully" indirect methods (right). The direct approach essentially involves first discretizing and then optimizing, whereas the indirect approach involves first applying a first-order optimality condition and then discretizing the optimality system. However, there are a number of "hybrid" intermediate possibilities. All of the methods illustrated in this figure are explained in detail in the book with the help of carefully chosen examples. Readers can therefore use Fig. 1 as a guide to the various approaches explained in the book.

As all the approaches mentioned above concern differentiable optimization methods, we present a free-derivative optimization method in the appendix. This method is included in `FreeFEM`, and we illustrate its use on a parameter identification problem.

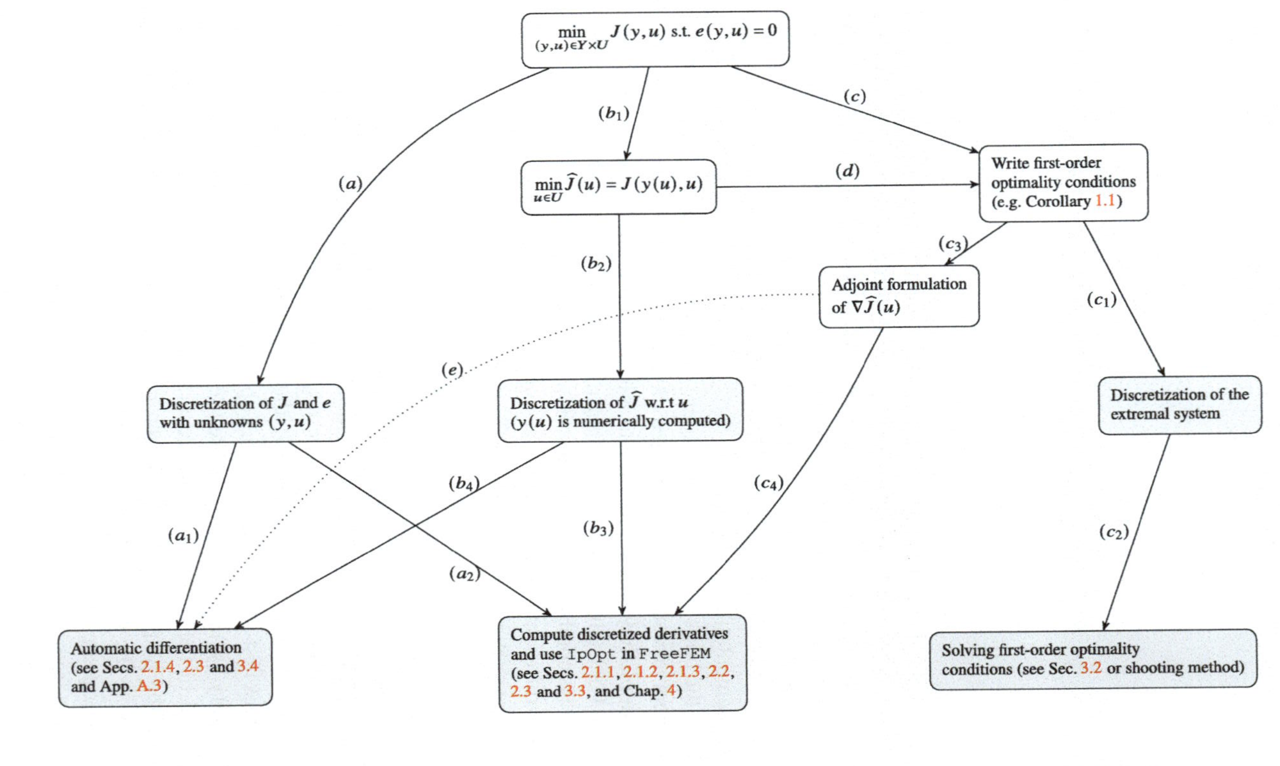

Fig. 1 Summary diagram of the differentiable optimization methods

It is important to emphasize the need for a well-defined theoretical framework within which the problem under consideration is well posed, before being discretized for numerical solution. Indeed, the success of the subsequent numerical implementation depends on the correct formulation. This is why, in the first part of the book, we review some fundamental principles of optimal control theory for partial differential equations, along with the optimization and discretization methods employed throughout. Then, we move on to the specifics of the various implementations.

Chapter 1 is devoted to introducing the framework and main tools that will be used throughout the book in the context of PDE-constrained optimization with `FreeFEM` and `IpOpt`. Chapter 2 discusses the linear quadratic case and its extension to the time-dependent case. This is followed by brief developments on the semilinear case and optimal shape design with relaxation. Chapter 3 has the dual aim of illustrating the numerical resolution of a calculus of variations problem and demonstrating the impact of choosing "*discretize then optimize*" or "*optimize then discretize*" on the numerical solution. Chapter 4 has been written to demonstrate how Chaps. 1 and 2 can be used to find the numerical solution to a challenging problem. We address the issue of optimally designing a micro-swimmer by formulating the problem within a practical framework and applying the aforementioned method. Appendix A includes several extras, such as the possible use of `FreeFEM` and `IpOpt` in Python. Appendix B illustrates `CMA-ES` as a possible alternative to `IpOpt` for a parameter identification problem.

We hope that readers will find this book useful and be able to use the provided models and templates to solve their own PDE-constrained optimization problems quickly and efficiently.

Paris, France
Solaize, France
Paris, France
November 2025

Frédéric Hecht
Gontran Lance
Emmanuel Trélat

Acknowledgments We warmly thank all reviewers whose insightful comments and sound advice substantially contributed to shaping the educational book we aimed to produce. We are grateful to Olivier Pironneau for his careful proofreading and for the valuable discussions throughout the development of the manuscript, and to Grégoire Allaire for his constructive feedback. We owe particular thanks to Mariano Mateos, whose extensive proofreading and thorough examination of both the text and the accompanying codes significantly raised the overall quality of this work, well beyond our initial expectations.

Competing Interests The authors have no competing interests to declare that are relevant to the content of this manuscript.

Contents

Codes Available on FreeFEM's Website https://freefem.org/Optim/

beam_paramater_identification.edp Find the Lamé's parameters of a beam by the stochastic method CMA-ES embedded in FreeFEM. xvi, 107, 110

ffad.{dat,mod,run} Solve a stationary linear quadratic optimal control problem with automatic differentiation in AMPL. xv, 48

ipoptfreefem.edp Short example of IpOpt use in FreeFEM generating a general template. xv, 24

lq_stationary_AMPL.edp Prepare matrices needed for solving a stationary linear quadratic optimal control problem with automatic differentiation in AMPL. 47

lq_stationary_O1.edp Solve a stationary linear quadratic optimal control problem with FreeFEM and IpOpt through a sequential direct method. xv, 39

lq_stationary_O2.edp Solve a stationary linear quadratic optimal control problem with FreeFEM and IpOpt by a simultaneous direct method. xv, 39

lq_stationary_O2_hotrestart.edp Solve a stationary linear quadratic optimal control problem with FreeFEM and IpOpt by a simultaneous direct method, using hotrestart technique to solve the problem on finer mesh by initializing the algorithm with a previously computed solution on a rougher mesh. 98

lq_stationary_casadi.edp Solve a linear quadratic optimal control problem with automatic differentiation in CasADi. xvi, 103

lq_stationary_indirect.edp Solve a stationary linear quadratic optimal control problem with FreeFEM and IpOpt by an indirect method. xv, 44

lq_time_O1.edp Solve a time-dependent linear quadratic optimal control problem with FreeFEM and IpOpt by a sequential method. 52

lq_time_O2.edp Solve a time-dependent linear quadratic optimal control problem with FreeFEM and IpOpt by a simultaneous method. 52

lq_time_t_as_z.edp Solve a time-dependent linear quadratic optimal control problem with FreeFEM and IpOpt regarding the 2D problem as a 3D problem with time t as the third space variable z. xv, 52

minsurf.{dat,mod,run} Find a minimal area surface lying on a fixed boundary with automatic differentiation in AMPL. xvi, 72

minsurf_Ipopt.edp Find a minimal area surface lying on a fixed boundary with FreeFEM and IpOpt. xvi, 70

minsurf_fixedpoint_Newton.edp Find a minimal area surface lying on a fixed boundary by a combination of fixed point and Newton methods. xvi, 68

semilinear_fixedpoint+Newton_ipopt.edp Solve a semilinear quadratic optimal control problem with FreeFEM and IpOpt after solving the PDE with a combination of a fixed point method and a Newton method. xv, 56

semilinear_fixedpoint_ipopt.edp Solve a semilinear quadratic optimal control problem with FreeFEM and IpOpt after solving the PDE with a fixed point method. xv, 55

semilinear_fixedpoint_ipopt_with_less_computations.edp Solve a semilinear optimal control problem with FreeFEM and IpOpt after solving the PDE with an iterative point method for the state equations and avoiding to solve too many times this equation to be solved in the gradient computation. 57

shape_microswimmer.edp Find the optimal shape of a micro swimmer whose outline is assimilated to a moving boundary with FreeFEM and IpOpt. xvi, 79, 92, 97, 98

test_poisson_L.edp Solving a Poisson equation on a L-meshed domain. xv, 5

Listings

Chapter 1
State of the Art

Abstract We devote this first chapter to recalling the foundations for theoretical and numerical PDE-constrained optimization. First, we introduce the `FreeFEM` software and the finite element method for numerically solving the PDE constraint. We then present the theoretical framework for optimal control of PDEs and highlight the need for an adjoint variable to compute the derivative of the objective function. We then focus on the interior point method `IpOpt` and show with a simple example how it can be called from `FreeFEM`. Finally, we discuss different discretization strategies, distinguishing between the choices *first optimize then discretize* and *first discretize then optimize*, also showing how to use automatic differentiation.

1.1 Introduction

Engineering problems or theoretical research problems often lead to optimization problems governed by partial differential equations (PDEs). Advances in computational capabilities, PDE solvers and optimization algorithms have provided accurate and efficient methods for solving challenging PDE constrained optimization problems.

When considering an optimal control problem governed by partial differential equations, as we will show in the illustrative examples provided in this survey, it is particularly important to establish preliminarily a rigorous mathematical framework in which the problem is well posed, before deriving and designing appropriate numerical methods to solve it efficiently.

Throughout Chap. 2, we refer to [11, 15, 21, 22] regarding classical issues on PDE control theory and general optimization. There exist many classical methods for solving PDEs, such as finite differences, finite volumes, spectral methods and general Galerkin methods. Here we will specifically focus on the Finite Element Method (FEM) and we refer to [18] for variational formulations of PDE problems. This chapter can serve as an introduction to numerical PDE optimization; some basic knowledge is required in numerical computations as well as some basics in some programming language. We also advise the reader to have some basics in the finite element method as well as in optimization.

F. Hecht et al., *PDE-constrained optimization within FreeFEM*, SpringerBriefs on PDEs and Data Science, https://doi.org/10.1007/978-981-95-7049-2_1

Optimal control problems are optimization problems where the decision variables, called *controls*, are acting on the *state* variables through an ordinary differential equation (ODE) or a partial differential equation (PDE). Within this viewpoint, the control is the input and the resulting state is the output. The optimization problem consists in determining what is the best possible input over a class of possible inputs, so that the output satisfies some prescribed constraints and minimizes a given criterion.

Most often, the objective function depends on both state and control variables and thus requires, at least from the numerical point of view, to make explicit or to compute the state's dependence with respect to the control (i.e., the input-output mapping of the system). State y and control u are respectively assumed to belong to real Banach spaces Y and U. General optimal control problems considered in Chap. 2 are written in the abstract form

$$\min_{(y,u)\in Y\times U} J(y,u), \qquad e(y,u)=0, \qquad c(y,u)\in K,$$

where

$$J : Y \times U \mapsto \mathbb{R}$$

is the objective function,

$$e : Y \times U \mapsto Z$$

usually stands for some ODE or PDE constraint and

$$c : Y \times U \mapsto K$$

defines some additional state and/or control constraints. Here, Z is a real Banach space and K is a closed convex set. Well-posedness is assumed, meaning that, for every control $u \in U$ (input), the equation

$$e(y,u)=0$$

has a unique solution $y = y(u) \in Y$ (output). Establishing existence of solutions for such general optimization problems may happen to be a challenging issue, often using deep functional analysis and compactness arguments. Besides, uniqueness of the solution is often a consequence of strict convexity properties.

In this chapter, we do not report on existence and uniqueness issues. Our objective is to show how to compute numerically in an efficient way a solution whose existence has already been proved or at least assumed. Given an optimal control problem, we will emphasize, with several examples, the importance of having a rigorous mathematical framework in which not only is settled and solved the problem, but it is also a guide for designing appropriate discretizations of the problem and ensuring the convergence of the subsequent numerical method.

Throughout the book, we focus on finite element discretization methods, particularly in dimensions 1, 2 or 3. We present the software `FreeFEM` developed at *Laboratoire Jacques-Louis Lions* (Sorbonne Université), a PDE solver using the finite element method and based on variational formulations with user-friendly and powerful features. We will provide some examples and details on how to use it and we will also refer to the documentation [10] (also available online) where most features are described. Regarding optimization strategies, we use differentiable methods which require the computation of derivatives. `FreeFEM` is flexible enough to allow users to write their own optimization algorithm (e.g., gradient, BFGS, Newton methods), or to plug it into some existing optimization routines. We will focus particularly on the `IpOpt` routine (see [24]), an (open and free) expert optimization routine that is well adapted to large-scale nonlinear optimization problems, and we will show how to combine it with `FreeFEM`. In particular, we will explain how `IpOpt` can be called directly from `FreeFEM`. Since solving PDEs requires a number of variables that increases with the size of the mesh, we have to deal with a large number of state and control variables (more than a million in the usual problems) as well as many constraints. The options available in `IpOpt` guarantee a good adaptability and efficiency for general convex and non-convex optimization problems of large size. Nevertheless and in spite of their efficiency, differentiable methods are not always the best choice, this is why we add in Appendix B an example treated with a free derivation optimization method.

1.2 Preliminaries

The `FreeFEM` *Software*

`FreeFEM` (see [10]) is a software developed in `C++` to solve PDEs with the finite element method in dimensions 1, 2 or 3. The meshes are generated thanks to an advanced automatic mesh generator and most of usual finite element spaces are embedded in so that most of the PDEs considered in practice can be solved via the discretization of the associated variational formulation. The documentation contains a complete introduction for a quick start as well as many classical examples such as heat conduction, elasticity system, Navier-Stokes equations. Users who are not familiar with the language are invited to browse the section [1, Learning by examples] where multiple examples will allow to advance step by step. We will see that not only `FreeFEM` has a user-friendly interface but its syntax is also similar to the mathematical problems under consideration, thus facilitating greatly their transcription. Mathematically, the PDE problems considered must be solved in the context of an appropriate variational formulation (see, for example, [18]). Indeed, `FreeFEM` treats PDEs in their weak form, the variational formulation being discretized according to a well suited choice of finite elements. As a first basic example, let us consider the Poisson equation

$$-\Delta y = u \text{ in } \Omega, \qquad y \in H_0^1(\Omega), \tag{1.1}$$

in some bounded domain Ω with a Lipschitz boundary, for some u in $L^2(\Omega)$. Its variational formulation is:

$$\text{find } y \in H_0^1(\Omega), \quad \int_\Omega \nabla y \cdot \nabla v \, dx = \int_\Omega uv \, dx \qquad \forall v \in H_0^1(\Omega) \tag{1.2}$$

$$\text{i.e.,} \qquad a(y, v) - b(u, v) = 0, \qquad \forall v \in H_0^1(\Omega), \tag{1.3}$$

where

$$a(y, v) = (\nabla y, \nabla v)_{L^2(\Omega)}$$

is a continuous and coercive bilinear form and

$$b(u, \cdot) = (u, \cdot)_{L^2(\Omega)}$$

is a continuous linear form, both being defined on the Hilbert space $H_0^1(\Omega)$. The Lax-Milgram theorem guarantees the existence of a weak solution of (1.2).

In order to discretize (1.2), we introduce a triangulation of the domain Ω (i.e. a mesh) as well as an appropriate finite element space which guarantees that the finite dimensional linear system resulting from the discretization of (1.3) on the basis of the finite element space is well posed (i.e., boundedly invertible).

As an example, we solve (1.1) on a 2D L-shaped domain Ω (see Fig. 1.1). The script `FreeFEM` for generating the mesh is given in the code 1.1.

```
int[int] lab=[1,1,1,1];
mesh Th = square(20,20,label=lab);
Th=trunc(Th,x<0.5 | y<0.5, label=1);
```

Code 1.1 Mesh generation with boundary label equal to 1

The finite element space V_h is

$$V_h = \left\{ v \in H^1(\Omega), \quad \forall K \in T_h \ \ v_{|K} \in \mathbb{P}_1 \right\} = \text{Span}(\phi_i)_{i \in \{1..n_d\}}$$

where $\mathbb{P}_1$ is the space of continuous piecewise linear functions and (ϕ_i) is a basis of it. One of the noticeable advantages of `FreeFEM` is that the user does not need to encode the specificities of the mesh and the finite element functions. This is done automatically by the command **`fespace`**. The access to the various data of the mesh and the finite element functions is then direct. Moreover, we will see that the

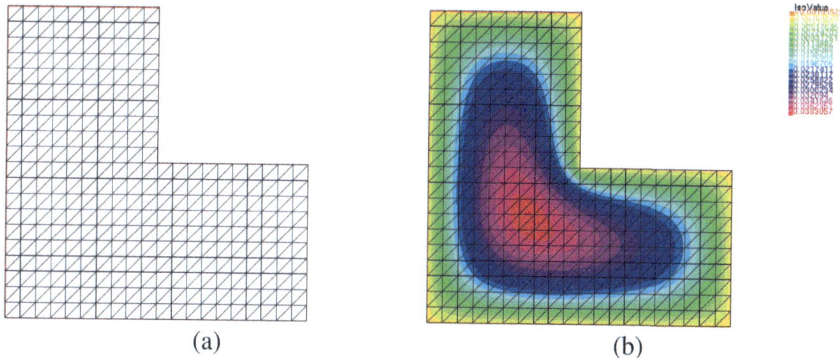

Fig. 1.1 Example (1.1): (**a**) L-mesh; (**b**) solution y of (1.2) for $u = 1$

use of macros allows to reduce significantly the number of lines of code to finally have a minimal script. The numerical solution of (1.2) is finally obtained by:

```
fespace Vh(Th,P1);
Vh Y,V,U=1; //finite element functions

macro grad(Y)  [dx(Y),dy(Y)] // //macro ended by //

solve Poisson(Y,V)  = int2d(Th)(grad(Y)'*grad(V))
                    - int2d(Th)(U*V)
                    + on(1,Y=0); // y ∈ H_0^1(Ω)
plot(Th,Y); // Fig.1.1
```

Code 1.2 Poisson solution by solving the variational formulation (see **test_poisson_L.edp**)

Instead of solving the variational form (1.2) with a single line of code, we can also define the matrices coming from the discretization of the bilinear form a and of the linear form b in (1.3), and solve the resulting linear system. We introduce the finite element subspace

$$Y_h = \{v \in V_h, \quad v_{|\partial\Omega} = 0\}$$

including the homogeneous Dirichlet boundary condition stated in (1.1) and let

$$(\phi_i)_{i\in\{1..n\}} \quad 0 < n < n_d,$$

be a basis of Y_h. The stiffness and mass matrices are respectively given by

$$(A_{h,ij})_{(i,j)\in\{1..n\}} = \int_\Omega \nabla\phi_i \cdot \nabla\phi_j \, dx, \qquad (M_{h,ij})_{(i,j)\in\{1..n_d\}} = \int_\Omega \phi_i\phi_j \, dx.$$

Solving (1.3) discretized by finite elements $\mathbb{P}_1$ finally consists in solving the linear system

$$A_h y_h - M_h u_h = 0.$$

Since the mesh T_h and the finite element space V_h are generated as in Code 1.1, `FreeFEM` is able to define the matrices A_h and M_h of the variational form (1.2) thanks to the command **`varf`** which is designed to construct variational forms in `FreeFEM`. It is not necessary to introduce the space of zero finite elements on the edge Y_h in the code. Indeed, `FreeFEM` handles the Dirichlet boundary conditions (homogeneous or inhomogeneous) by specifying them directly in the variational formulation.

```
Vh Y,V,U=1; // finite element functions

macro grad(Y) [dx(Y),dy(Y)] //
varf stiffness(Y,V) = int2d(Th)((grad(Y)'*grad(V)))
                        + on(1,Y=0); // y ∈ H_0^1(Ω)
varf Umass(Y,V) = int2d(Th)(U*V); // Y is not used. This is a way
    to define a linear form through the variational form tool

matrix Ah = stiffness(Vh,Vh,solver=sparsesolver); // with
    Dirichlet B.C.
real[int] Uh = Umass(0,Vh); // To specify Y=0
```

Code 1.3 **`varf`** command

Finally, **`solve`** `Poisson(Y,V)` in Code 1.2 is equivalent to solve $A_h y_h = u_h$ via

```
Y[] = Ah^-1*Uh;
```

Code 1.4 Poisson solution with finite element matrices

In both cases, it is not necessary to know the characteristics of the mesh in detail, which greatly facilitates the overall implementation. Note that, in Code 1.4, the notation `Ah^-1` does refer to the computation of the solution `Y[]` of the matrix equation `Ah Y[] = Uh`, and not to the computation of the inverse of the matrix.

Although it is close to writing the variational formulation, the **`solve`** command is most often not as fast as dealing directly with the matrices which are usually sparse. When we only deal with one PDE, it is not clear that one may save a significant computation time. But when in addition an optimal control problem has to be solved, the optimization process may require many calls to the state and adjoint PDE equations. There is then a genuine advantage to work with sparse matrices as soon as the number of PDEs considered increases. In Sect. 2.1, we will highlight the advantages of using sparse matrices to compute derivatives (in particular for Jacobian and Hessian matrices). Assuming that Eq. (1.1) is a constraint, we now want to minimize the functional J, depending on both y and u,

$$J(y, u) = \frac{1}{2}\int_{\Omega} (y - y_d)^2\, dx + \frac{\alpha}{2}\int_{\Omega} u^2\, dx, \tag{1.4}$$

whose computation requires solving (1.1) in a first step. Moreover, we will see that computing the derivative of J with respect to u may require to solve the state and adjoint PDEs coming from the first-order necessary optimality conditions, as explained in section "PDE-constrained Optimization" hereafter.

Let us mention some other softwares which are comparable to `FreeFEM` like NGsolve or Fenics. To go further with `FreeFEM` we refer the reader to a complete documentation on its website (see [1]). Main information, installation's guide, finite element method reminders, several examples, are available online.

PDE-constrained Optimization

In differentiable optimization problems, numerical methods are generally based on first-order optimality conditions and thus on the computation of derivatives. Let W be a Banach space, we denote its topological dual by $W' = \mathcal{L}(W, \mathbb{R})$ (space of continuous linear functionals on W) and by

$$\langle z, w\rangle_{W',W} = z(w) \quad \forall z \in W'$$

the duality pairing. Given a linear operator A, its adjoint is denoted by A^*. For the special case of the Hilbert spacc H, the dual H' can be directly identified with H and the duality pairing is identified with the inner product $(\cdot, \cdot)_H$ of H. A differentiable optimization problem is generally written as

$$\min_{w \in \mathcal{U}_{ad}} J(w) \tag{1.5}$$

where $J : W \to \mathbb{R}$ is Gateaux differentiable and $\mathcal{U}_{ad} \subset W$ is a nonempty, closed and convex subset. We denote by DJ the (Gateaux) derivative of J. First-order optimality conditions are then written in the following form.

Theorem 1.1 *Let $\bar{w} \in W$ be a (locally) optimal solution of* (1.5). *Then*

$$\bar{w} \in \mathcal{U}_{ad}, \quad \langle DJ(\bar{w}), w - \bar{w}\rangle_{W',W} \geqslant 0, \qquad \forall w \in \mathcal{U}_{ad}. \tag{1.6}$$

Classical differentiable optimization strategies involve the computation of (at least) first-order derivatives. Indeed, a Taylor expansion returns at some iterate point x_k

$$J(x) = J(x_k) + \langle DJ(x_k), x - x_k \rangle_{W',W} + o(\|x - x_k\|_W)$$

and the next iterate x_{k+1} is searched such that

$$J(x_{k+1}) \leqslant J(x_k)$$

and thus, assuming that x_{k+1} is close enough to x_k, such that

$$\langle DJ(x_k), x - x_k \rangle_{W',W} \leqslant 0.$$

In the case where $W = \mathbb{R}^N$, a particular descent direction is usually given by the opposite of the gradient $-\nabla J(x_k)$ (see Algorithm 1.1). However, the framework provided by (1.5) and Theorem 1.1 is rather limited when one has to tackle an optimal control problem because the objective function and its derivative are generally not easy to compute. Moreover, such an elementary algorithm may not perform well for large-scale optimization problems involving a PDE constraint. A new framework is therefore needed with the first constraint of decoupling state and control variables. Then, assuming the existence of solutions, the computation of the derivatives of the involved functions should be easy and we should be able to move easily to a powerful numerical framework. We therefore write below the general minimization problem of a function J depending on both *state* and *control* variables subject to a PDE embedded in the operator e with some additional constraints encoded in $\mathcal{U}_{ad}$:

$$\min_{(y,u)\in Y\times U} J(y,u) \quad \text{subject to} \quad e(y,u) = 0, \quad u \in \mathcal{U}_{ad}. \tag{1.7}$$

Algorithm 1.1 Gradient descent algorithm

initialization x_0, stop criterion ϵ
while $\|\nabla J(x_k)\| \geqslant \epsilon$ **do**
  compute α_k with linear search methods (in the direction $-\nabla J(x_k)$)
  compute $x_{k+1} = x_k - \alpha_k \nabla J(x_k)$
  compute $\nabla J(x_{k+1})$
end while

Remark 1.1 In a more general way, we can consider in addition to the equality constraints, inequality constraints like

$$g(y,u) \leqslant 0,$$

as it is the case for example in obstacle problems. In the following, we only focus on equality constraints, noting that inequality constraints are in general easier to treat than equality constraints by direct methods.

Following [11], the assumptions 1.2 below guarantee existence of a solution $(\bar{y}, \bar{u})$ in $Y \times U$ of (1.7).

Assumption 1.2

1. $\mathcal{U}_{ad} \subset U$ is nonempty, closed and convex.
2. The mappings

$$J : Y \times U \mapsto \mathbb{R} \quad \text{and} \quad e : Y \times U \mapsto Z$$

   are continuous, where Z is a Banach space and Y, U are reflexive Banach spaces.
3. For every $u \in \mathcal{U}_{ad}$, the state equation

$$e(y, u) = 0$$

   has a unique solution $y(u) \in Y$ and the mapping

$$u \in \mathcal{U}_{ad} \mapsto y(u) \in Z$$

   is continuous.
4. The mapping

$$(y, u) \in Y \times U \mapsto e(y, u) \in Z$$

   is weakly continuous.
5. J is sequentially weakly lower-semicontinuous.

Some state constraints can be added by means of a set $\mathcal{Y}_{ad} \subset Y$ assumed to be nonempty, convex and closed. We introduce the reduced cost function of the problem (1.7)

$$u \in \mathcal{U}_{ad} \mapsto \widehat{J}(u) = J(y(u), u)$$

so that (1.7) is reformulated as

$$\min_{u \in \mathcal{U}_{ad}} \widehat{J}(u).$$

If we want to use Algorithm 1.1 to compute an optimal solution, we have to compute the derivative of $\widehat{J}$. This requires to compute the derivative of the mapping $u \in \mathcal{U}_{ad} \mapsto y(u) \in Y$, which is not explicit. Assumption 1.3 below provides a general framework that ensures the differentiability of the input-output mapping $u \in \mathcal{U}_{ad} \mapsto y(u) \in Y$ (by the implicit function theorem) and at the same time allows us to compute $D\widehat{J}$, which is necessary to derive the first-order optimality conditions.

Assumption 1.3

1. $\mathcal{U}_{ad} \subset U$ is nonempty, closed and convex.
2. The mappings

$$J : Y \times U \to \mathbb{R} \quad \text{and} \quad e : Y \times U \to Z$$

   are continuously Fréchet differentiable, where U, Y and Z are Banach spaces.
3. For all $u \in V$ in a neighborhood V of $\mathcal{U}_{ad}$, the state equation

$$e(y, u) = 0$$

   has a unique solution $y(u) \in Y$.
4. The partial derivative

$$\partial_y e(y(u), u) \in \mathcal{L}(Y, Z)$$

   has a bounded inverse for every $u \in \mathcal{U}_{ad}$.

Applying Theorem 1.1 to $u \in \mathcal{U}_{ad} \mapsto \widehat{J}(u)$, we get the following first-order optimality conditions in terms of the reduced cost function $\widehat{J}$.

Theorem 1.4 *Under Assumption 1.3, if $\widehat{u} \in \mathcal{U}_{ad}$ is a (locally) optimal solution of the reduced problem, then*

$$\langle D\widehat{J}(\widehat{u}), u - \widehat{u} \rangle_{U',U} \geqslant 0 \qquad \forall u \in \mathcal{U}_{ad}.$$

At this step, a descent direction can be found by exploiting the derivative $D\widehat{J}$. Unfortunately, Theorem 1.4 does not provide an easy way to compute it numerically since, according to the sensitivity analysis developed below, the numerical computation of the derivative of the mapping $u \in \mathcal{U}_{ad} \mapsto y(u) \in Y$ requires the computation of "too many" directional derivatives.

Sensitivity Approach

Let s be such that $u + \epsilon s \in \mathcal{U}_{ad}$ for ϵ small enough and then compute $\widehat{J}(u + \epsilon s)$. Thanks to Assumption 1.3, the chain rule gives

$$\langle D\widehat{J}(u), s\rangle_{U',U} = \left\langle \partial_y J(y(u), u), Dy(u)s\right\rangle_{Y',Y} + \left\langle \partial_u J(y(u), u), s\right\rangle_{U',U}. \quad (1.8)$$

The partial derivatives $\partial_y J$ and $\partial_u J$ are easy to compute since J explicitly depends on y and u. In contrast, computing $Dy(u)s$ requires to solve $e(y, u) = 0$ and this is not straightforward. Differentiating

$$e(y(u), u) = 0$$

in the direction s allows us to compute $\delta y_s = Dy(u)s$ as the solution of the linear equation

$$\partial_y e(y(u), u)\delta y_s = -\partial_u e(y(u), u)s. \quad (1.9)$$

Computing

$$\langle D\widehat{J}(u), s\rangle_{U',U}$$

thus requires to compute the solution δy_s of (1.9) for each direction s. Hence, the numerical computation of the differential $D\widehat{J}(u)$ is difficult if U has a large dimension since it is necessary to compute the directional derivative in each direction of a given basis of the vector space spanned by U. A numerical method based on this sensitivity approach is therefore not feasible in high dimension because it would be too much computationally demanding. The same problem is encountered for automatic differentiation in the direct mode (as explained in section "Automatic Differentiation"), where the Jacobian is not necessarily needed because we generally need a descent direction that is given by the Jacobian applied to a well-chosen direction. In the same way as automatic differentiation does in the reverse mode (see [12]), the gradient can be found with significantly less efforts by introducing an adjoint variable, as explained next.

Adjoint Approach

Equation (1.8) is equivalently written as

$$\langle D\widehat{J}(u), s\rangle_{U',U} = \left\langle Dy(u)^* \partial_y J(y(u), u), s\right\rangle_{U',U} + \left\langle \partial_u J(y(u), u), s\right\rangle_{U',U}$$

and thus

$$D\widehat{J}(u) = Dy(u)^* \partial_y J(y(u), u) + \partial_u J(y(u), u).$$

Moreover, (1.9) gives

$$\partial_y e(y(u), u) Dy(u) = -\partial_u e(y(u), u). \tag{1.10}$$

It is not required to know the whole differential $Dy(u)$ but only the vector

$$Dy(u)^* \partial_y J(y(u), u).$$

The fourth item of Assumption 1.3 ensures the existence of the inverse $-\partial_y e(y(u), u)^{-1}$ and (1.10) gives

$$Dy(u) = -\partial_y e(y(u), u)^{-1} \partial_u e(y(u), u)$$

and

$$\begin{aligned} Dy(u)^* \partial_y J(y(u), u) &= \Big(- \partial_y e(y(u), u)^{-1} \partial_u e(y(u), u) \Big)^* \partial_y J(y(u), u) \\ &= -\partial_u e(y(u), u)^* \underbrace{\left(\partial_y e(y(u), u)^*\right)^{-1} \partial_y J(y(u), u)}_{\text{adjoint}} - p(u). \end{aligned}$$

The adjoint vector $p = p(u) \in Z'$ is defined as the solution of the linear equation

$$\partial_y e(y(u), u)^* p = -\partial_y J(y(u), u). \tag{1.11}$$

Finally,

$$D\widehat{J}(u) = \partial_u e(y(u), u)^* p(u) + \partial_u J(y(u), u).$$

From a numerical point of view, compared with the sensitivity approach which requires to solve as many PDEs as U has degrees of freedom to express the derivative of $\widehat{J}$, the adjoint approach only requires to solve the state equation in (1.7) and the adjoint equation (1.11). This gives a significant advantage in that solving the PDEs requires more time and more computation as the mesh is finer. Hence, the computation of the first derivative of the objective function of the problem (1.7) at a point $u \in \mathcal{U}_{ad}$ follows the following steps:

S.1 Compute the partial derivatives

$$\partial_y J(y, u), \quad \partial_u J(y, u), \quad \partial_y e(y, u), \quad \partial_u e(y, u),$$

and the adjoint operators

$$\partial_y e(y,u)^*, \quad \partial_u e(y,u)^*.$$

S.2 Solve the state equation

$$e(y,u) = 0,$$

which gives $y(u)$ (input-output mapping).

S.3 Solve the adjoint equation

$$\partial_y e(y(u),u)^* p = -\partial_y J(y(u),u),$$

which gives the adjoint $p = p(u)$.

S.4 Finally, compute

$$D\widehat{J}(u) = \partial_u e(y(u),u)^* p(u) + \partial_u J(y(u),u).$$

The adjoint variable p can be interpreted as the Lagrange multiplier corresponding to the constraint $e(y,u) = 0$. The Lagrangian $L : Y \times U \times Z' \to \mathbb{R}$ of the problem (1.7) is defined by

$$L(y,u,p) = J(y,u) + \langle p, e(y,u)\rangle_{Z',Z}\,.$$

The partial derivatives of the Lagrangian with respect to the adjoint variable p and the state variable y give the state equation in (1.7) and the adjoint equation (1.11), respectively, while the partial derivative with respect to the control variable u gives the derivative of the reduced cost function $\widehat{J}$. Under Assumption 1.3, Theorem 1.4 is reformulated as follows (Pontryagin maximum principle, well known in optimal control theory).

Corollary 1.1 *Under Assumption 1.3, let* $(\bar{y},\bar{u}) \in Y \times U$ *be an optimal solution of* (1.7) *with* $\bar{u} \in \mathcal{U}_{ad}$*. Then there exists* $\bar{p} \in Z'$ *such that*

$$e(\bar{y},\bar{u}) = 0, \tag{1.12}$$

$$\partial_y e(\bar{y},\bar{u})^* \bar{p} = -\partial_y J(\bar{y},\bar{u}), \tag{1.13}$$

$$\big\langle \partial_u J(\bar{y},\bar{u}) + \partial_u e(\bar{y},\bar{u})^* \bar{p}, u - \bar{u}\big\rangle_{U',U} \geqslant 0 \quad \forall u \in \mathcal{U}_{ad}. \tag{1.14}$$

(continued)

Corollary 1.1 (continued)
The Lagrangian formulation of the optimality conditions is

$$\langle q, \partial_p L(\bar{y}, \bar{u}, \bar{p})\rangle_{Z',Z} = 0 \qquad \forall q \in Z', \tag{1.15}$$

$$\langle \partial_y L(\bar{y}, \bar{u}, \bar{p}), v\rangle_{Y',Y} = 0 \qquad \forall v \in Y, \tag{1.16}$$

$$\langle \partial_u L(\bar{y}, \bar{u}, \bar{p}), u - \bar{u}\rangle_{U',U} \geqslant 0 \qquad \forall u \in \mathcal{U}_{ad}. \tag{1.17}$$

Corollary 1.1 is most often used to find the adjoint equation and the derivative of $\widehat{J}$. This requires to identify precisely the involved spaces Y, U and Z and the duality pairings must be chosen accordingly. When U is a Hilbert space, $D\widehat{J}(u)$ can be identified with the gradient $\nabla\widehat{J}(u)$ corresponding to the inner product of U.

Remark 1.2 Corollary 1.1 provides necessary first-order optimality conditions for a PDE constrained optimization problem. Such conditions are known to be sufficient when the problem (1.7) is convex. Otherwise, one can use the sufficient conditions for local optimality given by second-order optimality conditions (and the Hessian). However, from a numerical point of view, the numerical computation of the Hessian of the Lagrangian might be heavy (the matrix may not be sparse) and can be replaced by an approximated matrix (BFGS and quasi-Newton method for instance). Sometimes, we also can circumvent this problem by computing the Hessian applied to an appropriate direction since the whole matrix is not useful (see, e.g., [19]).

Remark 1.3 The introduction of the adjoint is an efficient way to compute the derivative of the objective function. The additional control constraints included in the set $\mathcal{U}_{ad}$ can also be handled. However, it is much more difficult to take into account the possible additional state constraints included in $\mathcal{Y}_{ad}$ because this requires to modify accordingly the adjoint equation, which may become more difficult to solve or even be a challenge to write. In such a case, the adjoint method may not be advisable. We will give some alternatives in Chap. 2.

Poisson Example

Let Ω be an open subset of $\mathbb{R}^N$ with Lipschitz boundary. We have defined in section "The `FreeFEM` Software" the problem of minimizing the cost function (1.4) subject to a Poisson PDE with homogeneous Dirichlet boundary condition (1.1), that we reformulate in the framework of (1.7) by introducing the sets $Y = H_0^1(\Omega)$ and $U = L^2(\Omega)$. For the moment, we do not consider any additional control constraints included in the set $\mathcal{U}_{ad} = U$. We thus denote the cost function and PDE constraints by the functions J and e in a weak form:

$$J : (y, u) \in Y \times U \mapsto \frac{1}{2}\int_\Omega (y - y_d)^2\,dx + \frac{\alpha}{2}\int_\Omega u^2\,dx \in \mathbb{R}$$

$$e : (y, u) \in Y \times U \mapsto a(y, \cdot) - b(u, \cdot) \in Z$$

where a and b are the bilinear forms defined by (1.3). Consider the Gelfand triple (see [11, Definition 1.26])

$$H_0^1(\Omega) \subset L^2(\Omega) \subset H^{-1}(\Omega),$$

where $L^2(\Omega)$ is identified with its dual so that the duality pairing $\langle\cdot,\cdot\rangle_{Y',Y}$ is compatible with the $L^2(\Omega)$-inner product. Since $U = L^2(\Omega)$, the duality pairing $\langle\cdot,\cdot\rangle_{U',U}$ is the $L^2(\Omega)$-inner product. Finally, we set $Z = H^{-1}(\Omega) = Y'$ so that the duality pairing $\langle\cdot,\cdot\rangle_{Z',Z}$ is compatible with the $L^2(\Omega)$-inner product. Assumption 1.3 is verified. Indeed, Items *1.* and *2.* are straightforward while Items *3.* and *4.* are due to properties of elliptic operators stated in [7]. For $p \in Z' = H_0^1(\Omega)$, the Lagrangian is given by

$$L(y, u, p) = \int_\Omega \left(\frac{1}{2}(y - y_d)^2 + \frac{\alpha}{2}u^2 + \nabla y \cdot \nabla p - up\right) dx.$$

Remark 1.4 The identification of the duality pairings induced by the $L^2(\Omega)$-inner product gives a Lagrangian that we can handle easily. Nevertheless, this compatibility depends on the sets Y, Z and U chosen to verify Assumption 1.3 and is not always straightforward to obtain.

We finally apply Corollary 1.1 to express the weak formulation of the adjoint equation and exhibit the variational inequality which gives the first derivative of the reduced cost function $\widehat{J}$ that is then identified with the gradient associated with the $L^2(\Omega)$-inner product.

$$\partial_p L(y, u, p) = 0 \iff \int_\Omega (\nabla y \cdot \nabla v - uv)\,dx = 0 \quad \forall v \in H_0^1(\Omega),$$

$$\partial_y L(y, u, p) = 0 \iff \int_\Omega (\nabla p \cdot \nabla v + (y - y_d)v)\,dx = 0 \quad \forall v \in H_0^1(\Omega),$$

$$\partial_u L(y, u, p) : v \in L^2(\Omega) \mapsto \int_\Omega (\alpha u - p)v\,dx,$$

i.e., in the strong form for the $L^2(\Omega)$-inner product,

$$\begin{aligned} y \in H_0^1(\Omega) \text{ solution of } \; -\Delta y &= u \quad \text{in } \Omega, \\ p \in H_0^1(\Omega) \text{ solution of } \quad \Delta p &= y - y_d \quad \text{in } \Omega, \\ D\widehat{J}(u) \text{ identified with } \nabla \widehat{J}(u) &= \alpha u - p. \end{aligned}$$

Dirichlet Boundary Control Example

In this section, we focus on a boundary control problem. We further assume that Ω has either a C^2 boundary or is a convex polytope. In the previous example, the variational formulation was written by introducing the set $Y = H_0^1(\Omega)$ for a homogeneous Dirichlet boundary condition. For Neumann or Robin boundary conditions, we take $Y = H^1(\Omega)$ instead. For inhomogeneous Dirichlet boundary conditions, the standard variational formulation must be reformulated using some alternatives.

Given $f \in L^2(\Omega)$, we modify the previous example by adding a Dirichlet boundary condition $u \in L^2(\partial\Omega)$ so that the new problem is

$$\min J(y, u) = \frac{1}{2} \int_\Omega (y(x) - y_d(x))^2 \, dx + \frac{\alpha}{2} \int_{\partial\Omega} u(x)^2 \, dx \tag{1.18}$$

$$\text{subject to} \begin{cases} -\Delta y = f & \text{in } \Omega, \\ y = u & \text{on } \partial\Omega. \end{cases} \tag{1.19}$$

We cannot write directly the weak formulation as in the Poisson example. To overcome this difficulty, a possibility may be to first introduce a small parameter δ so that the Dirichlet boundary condition in (1.19) becomes a Robin boundary condition

$$\delta \partial_n y + y = u \text{ on } \partial\Omega$$

and to write the variational formulation by introducing the space $Y = H^1(\Omega)$. Here, we consider instead the way `FreeFEM` handles Dirichlet boundary conditions. Following [23, Sec. 10.6], we denote by

$$A_0 : \mathcal{D}(A_0) \to L^2(\Omega)$$

the Dirichlet Laplacian (we have $\mathcal{D}(A_0) = H_0^1(\Omega) \cap H^2(\Omega)$ because of the regularity assumption on Ω) and by γ_0, γ_1 respectively the *Dirichlet* and *Neumann* traces. The Dirichlet map $\mathbf{D}$ is defined as follows: for any $u \in L^2(\partial\Omega)$ there exists a unique $\mathbf{D}u \in L^2(\Omega)$ such that

$$\Delta \mathbf{D}u = 0 \text{ on } \Omega \quad \text{and} \quad \gamma_0(\mathbf{D}u) = \mathbf{D}u_{|\partial\Omega} = u$$

(actually, $\mathbf{D}u \in C^\infty(\Omega)$ and the operator $\mathbf{D}$ is bounded from $L^2(\partial\Omega)$ to $L^2(\Omega)$). The adjoint of $\mathbf{D}$ is

$$\mathbf{D}^* = -\gamma_1 A_0^{-1}$$

and for all $v \in L^2(\Omega)$

$$(\mathbf{Du}, v)_{L^2(\Omega)} = -(u, \partial_n\phi)_{L^2(\partial\Omega)} \text{ with } A_0\phi = v.$$

We thus seek the solution $y(u)$ of (1.19) in the affine space $H_0^1(\Omega) + \mathbf{D}u$ and we define $z \in H_0^1(\Omega) \cap H^2(\Omega)$ the solution of

$$\begin{aligned} -\Delta z &= f \quad \text{in } \Omega \\ z &= 0 \quad \text{on } \partial\Omega, \end{aligned}$$

so that $y = z + \mathbf{D}u$ with $z = A_0^{-1} f$, whose variational formulation is the following: find $z \in H_0^1(\Omega)$ such that

$$\int_\Omega \nabla z \cdot \nabla v = \int_\Omega f v\, dx \quad \forall v \in H_0^1(\Omega).$$

The state space Y has to be defined in order to find a unique weak solution $y \in Y$ of (1.19) when solving the resulting *very weak* variational formulation

$$\int_\Omega -y\Delta v\, dx = \int_\Omega f v\, dx - \int_\Omega \mathbf{D}u \Delta v\, dx \quad \forall v \in H_0^1(\Omega) \cap H^2(\Omega).$$

Since the Dirichlet Laplacian A_0 induces an isomorphism from $H_0^1(\Omega) \cap H^2(\Omega)$ to $L^2(\Omega)$ and A_0^{-1} is also selfadjoint in $L^2(\Omega)$, one can take $\phi \in L^2(\Omega)$ such that

$$v = A_0^{-1}\phi$$

and the previous variational formulation is equivalent to:

$$\text{find } y \in Y \text{ s.t.} \quad \int_\Omega y\phi\, dx = \int_\Omega \left(A_0^{-1} f\right) \phi\, dx + \int_\Omega \mathbf{D}u\phi\, dx \qquad \forall \phi \in L^2(\Omega).$$

Therefore, we set $Y = L^2(\Omega)$, $U = L^2(\Gamma)$ and $Z = L^2(\Omega)$ so that

$$y = z + \mathbf{D}u \subset L^2(\Omega)$$

is the unique solution of (1.19) (uniqueness of solution is straightforward by putting $(f, u) = 0$). Then we define

$$e(y, u) = y - A_0^{-1} f - \mathbf{D}u$$

so that Assumption 1.3 is satisfied in that setting and optimality conditions yield the existence of $\phi \in L^2(\Omega)$ such that

$$\phi = y_d - y$$
$$\left(\mathbf{D}^*\phi, v - u\right)_{L^2(\partial\Omega)} \geqslant 0 \quad \forall v \in U.$$

Introducing $p \in H_0^1(\Omega) \cap H^2(\Omega)$ solution of $p = A_0^{-1}\phi$, the adjoint equation now reads

$$\Delta p = y - y_d \quad \text{in } \Omega$$
$$p = 0 \qquad \text{on } \partial\Omega,$$

so that $D\widehat{J}(u)$ is identified with the gradient

$$\nabla \widehat{J}(u) = \alpha u + \partial_n p$$

for the $L^2(\partial\Omega)$-inner product.

Remark 1.5 If none of the assumptions done on Ω hold, we have to modify the space $H_0^1(\Omega) \cap H^2(\Omega)$ accordingly, since existence and uniqueness of the solution to the very weak variational formulation relies on the isomorphism $A_0 \in \mathcal{L}\left(H_0^1(\Omega) \cap H^2(\Omega), L^2(\Omega)\right)$. This issue is treated in [23, Sec. 13].

Remark 1.6 Regarding the solution of (1.19), given $\mathcal{U}_{ad} = \{u \in L^2(\partial\Omega), a \leqslant u(x) \leqslant b, a.e. x \in \partial\Omega\}$, it is usually the case for y to be $H^1(\Omega)$. We refer to [4] where an elusive counterexample has been found. In such a case, we solve (1.19) in `FreeFEM` by searching the solution y in $\{w \in H^1(\Omega), w_{|\partial\Omega} = u\}$ verifying

$$\int_\Omega \nabla y \cdot \nabla v \, dx = \int_\Omega f v \, dx \qquad \forall v \in H_0^1(\Omega).$$

This is numerically carried out by specifying the boundary conditions thanks to the command `on(IndexBoard, Y=U)`.

```
Vh Y,V; // finite element functions
macro grad(Y) [dx(Y),dy(Y)] //
solve Poisson(Y,V) = int2d(Th)((grad(Y)'*grad(V)))
                   + on(1,Y=U); // y ∈ {w ∈ H^1(Ω), w_|∂Ω = u}
```

We mention this inhomogeneous Dirichlet limit problem to highlight the difficulties that can arise with the choice of Y, Z and U spaces to theoretically find the adjoint and derivative. From the numerical point of view, although we generally do not need to know these sets explicitly since we often assume that all involved data are sufficiently smooth, a good understanding of the mathematical framework and in particular the knowledge of the discretized spaces as well as the inner products is crucial in the calculations.

When we move to the numerical framework and have in mind the Algorithm 1.1, the numerical computation of the derivative of the objective function involves the adjoint variable found by solving the adjoint equation according to a chosen scheme. Alternatively, one can override the adjoint equation and directly derive a discretized version of the optimal control problem. Both approaches are respectively called *First Optimize Then Discretize (FOTD)* and *First Discretize Then Optimize (FDTO)*.

Overview of Optimization's Strategies

Given a general optimal control problem, we have given in section "PDE-constrained Optimization" ways to compute the continuous derivatives of the involved functions. This numerically implies to get a suitable approximation of both functions and of their derivatives that allow their numerical computation. Effectiveness of the procedure is directly imputed to handling both discretization and optimization. Given discretization parameters $0 < h < h_0$ and some families of finite-dimensional discretization spaces $(Y_h)_{0<h<h_0}$, $(U_h)_{0<h<h_0}$, $(Z_h)_{0<h<h_0}$ and $(\mathcal{U}^h_{ad})_{0<h<h_0}$ of the spaces Y, U, Z and $\mathcal{U}_{ad}$, the optimal control problem (1.7) is discretized as

$$\min_{(y_h,u_h)\in Y_h\times U_h} J_h(y_h,u_h) \quad \text{subject to:} \quad e_h(y_h,u_h)=0, \quad u_h\in\mathcal{U}^h_{ad}. \tag{1.20}$$

where

$$J_h : Y_h \times U_h \to \mathbb{R} \text{ and } e_h : Y_h \times U_h \to Z$$

are discretized versions of the functions J and e. Without loss of generality, the finite-dimensional spaces Y_h and U_h are identified with vector spaces $\mathbb{R}^{N_Y}$ and $\mathbb{R}^{N_U}$, with $N_Y \geqslant 1$, $N_U \geqslant 1$. The problem (1.20) is expressed in a finite-dimensional framework and can then be numerically solved by usual tools.

In this book, we particularly focus on differentiable methods, which therefore call, at least, the first-order derivatives of the cost and constraint functions. Nevertheless, we may encounter difficulties when the functions concerned are not smooth enough. Moreover, another difficulty appears when the cost function is not convex and has a real importance when the number of local extrema increases. This is why, in Appendix B, we will focus on a derivative-free optimization strategy for a parameter identification problem. Indeed, for such a problem, we are not sure to be able to differentiate functions with respect to the parameters involved in the PDE. Moreover, most identification problems require a large amount of experimental data which leads to write a cost function as a sum of several polynomial functions, thus with a large number of local extrema. Well known derivative-free methods are genetic algorithms, Bayesian methods in image processing problems and machine learning methods. In Appendix B, we will take advantage once again of `FreeFEM`, which benefits from the stochastic optimization routine `CMA-ES` (see [9]).

We now come back to differentiable optimization algorithms, which are directly related to the *KKT conditions* and need derivatives of the functions J_h and e_h. This approach is usually called *First Discretize Then Optimize (FDTO)* or *direct* method. In contrast, in the *First Optimize Then Discretize (FOTD)* (or *indirect*) approach, the first-order optimality condition of the continuous problem (1.7) is first derived by applying Corollary 1.1 such that all functions sets and operators involved are then discretized accordingly to get for an optimal solution $(\bar{y}_h, \bar{u}_h) \in Y_h \times \mathcal{U}_{ad}^h$

$$e_h(\bar{y}_h, \bar{u}_h) = 0, \tag{1.21}$$

$$(\partial_y e_h(\bar{y}_h, \bar{u}_h))^* p_h = -\partial_y J_h(\bar{y}_h, \bar{u}_h), \tag{1.22}$$

$$\left(\partial_u J_h(\bar{y}_h, \bar{u}_h) + (\partial_u e_h(\bar{y}_h, \bar{u}_h))^* \bar{p}_h, v_h - \bar{u}_h\right)_{U_h} \geqslant 0, \quad \forall v_h \in \mathcal{U}_{ad}^h. \tag{1.23}$$

Both methods may not be mathematically equivalent since the partial derivatives of the discretized functions J_h and e_h may differ from the discretizations of the partial derivatives $\partial_y e_h$, $\partial_u e_h$, $\partial_y J_h$, $\partial_u J_h$.

Note that the *FOTD* approach does not give the true numerical derivative of the discretized function (rather obtained with the *FDTO* approach). This may affect the algorithm's convergence.

To our knowledge, the question of the convergence of optimal solutions of the discretized problem to the solution of the continuous one is a challenging issue that deserves further consideration. On this issue, within the ODE context, under some appropriate assumptions on optimal control problems in finite dimension, it is proved in [20] that the direct *FDTO* and indirect *FOTD* approaches are mathematically equivalent when the discretization is performed with a *symplectic partitioned Runge-Kutta integrator*. This reference also contains interesting issues related to automatic differentiation, that we illustrate in Appendix A.2. Indeed, in this example, when the state equation is discretized according to an implicit scheme, automatic differentiation in reverse mode actually hides an explicit scheme for the adjoint equation.

When optimization is performed first, we have to pay a special attention to the choices of discretization for state and adjoint variables. Nevertheless the freedom of discretization's choice for the adjoint variable sometimes implies that *FOTD* is more relevant since the derivative produces the adjoint which is, without state constraints, usually more regular than the state variable. The adjoint can thus be approximated more accurately thanks to a well-fitting discretization. Conversely, one of the main advantages of *FDTO* approach is its ability to allow the use of large-scale optimization methods, state constraints and automatic differentiation. We mention [16, Section 6.2] for a comparison between the *FDTO* and *FOTD* approaches for a PDE-constrained optimization problem.

Remark 1.7 In this remark, we mention a general rigorous mathematical framework for discretizations [13, 14]. Under Assumption 1.3, let Y, U and Z be separable Banach spaces. Let $\tilde{Y}$ be a separable Banach space such that the embedding $Y \hookrightarrow \tilde{Y}$ is continous. Given a discretization parameter h and a family of finite-dimensional spaces $(Y_h)_{0<h<h_0}$ assumed to be uniformly continuously embedded $Y_h \hookrightarrow \tilde{Y}$, we assume the existence of projection and embedding operators

$$P_h^Y : \tilde{Y} \to Y_h \quad \text{and} \quad \tilde{P}_h^Y : Y_h \to \tilde{Y}$$

such that $P_h^Y \tilde{P}_h^Y = \mathrm{id}_{Y_h}$. The numerical scheme is assumed to be convergent, i.e.,

$$\lim_{h\to 0} \|\tilde{P}_h^Y P_h^Y y - y\|_{\tilde{Y}} = 0 \qquad \forall y \in Y.$$

Note that $\|\tilde{P}_h^Y\|_{\mathcal{L}(Y_h,\tilde{Y})} = 1$ and, by the Uniform Boundedness Principle, that $\|P_h^Y\|_{\mathcal{L}(\tilde{Y},Y_h)} \leqslant \mathrm{Cst}$ (uniform constant). The same setting is established for U and Z'. The mappings J and e involved in (1.7) are approximated by

$$J_h : (y_h, u_h) \in Y_h \times U_h \mapsto J(\tilde{P}_h^Y y_h, \tilde{P}_h^U u_h) \in \mathbb{R},$$
$$e_h : (y_h, u_h) \in Y_h \times U_h \mapsto e(\tilde{P}_h^Y y_h, \tilde{P}_h^U u_h) \in Z,$$

and, according to the *FOTD* approach, we have

$$\partial_\# e_h(y_h, u_h) = (\tilde{P}_h^{Z'})^* \partial_\# e(\tilde{P}_h^Y y_h, \tilde{P}_h^U u_h)\tilde{P}_h^\star,$$
$$\partial_\# J_h = \partial_\# J(\tilde{P}_h^Y y_h, \tilde{P}_h^U u_h)\tilde{P}_h^\star,$$

where $(\#, \star) \in \{(y, Y), (u, U)\}$. Such a framework encompasses most of the usual discretization strategies (finite differences, finite elements, Galerkin). The particular case of a linear PDE is addressed in [3].

For the Poisson example (1.1), having in mind its variational formulation (1.2), we take

$$Y = \tilde{Y} = H_0^1(\Omega), \quad U = \tilde{U} = L^2(\Omega), \quad Z' = \tilde{Z}' = H_0^1(\Omega)$$

so that $Y \hookrightarrow U \hookrightarrow Z$. The finite element space U_h consists of $\mathbb{P}_1$ Lagrange elements, Y_h is the subspace of U_h such that we have 0 on the boundary and $(Z_h)' = Y_h$. Embedding operators are thus induced by the above canonical embeddings and projections via the $L^2(\Omega)$-inner product.

A general framework for conformal transformations and especially Galerkin discretization methods in the context of PDE optimization is stated in [13, 14]. In the next page, we illustrate in a diagram several ways used throughout this survey for dealing with PDE constrained optimization. The corresponding codes are available on `FreeFEM`'s website.

On Fig. 1.2:

- (a) the continuous optimization problem is discretized according to well chosen discretization spaces for both state and control. The cost and the constraint (the PDE) functions are discretized accordingly to finally obtain an optimization problem in finite dimension.
- (a_1) We give the resulting NLP (Non Linear Programming) as input to a software equipped with an automatic differentiation tool and `IpOpt` (`AMPL` for instance). Then, derivatives and optimization are done by `AMPL`.
- (a_2) Or we directly compute derivatives of the previous optimization problem in finite dimension and directly call `IpOpt` in `FreeFEM`. At this step, derivatives returned by `AMPL` or by hand-computing should be identical.
- (b_1, b_2) As in step (a), we discretize the continuous optimization problem according to a well chosen discretization space for the control only. The cost function is discretized by first computing the discretized state with respect to the control. The PDE constraint is not anymore an explicit constraint but becomes step of the cost function computation.
- (b_3) As for steps (a_1, a_2), either we compute derivatives of the optimization problem in finite dimension (discretization of the reduce cost function) and give them as inputs to `IpOpt` in `FreeFEM`,
- (b_4) or we directly write the NLP problem in `AMPL` which proceeds on its own to the computation of derivatives and to the optimization process.
- (c) We write the continuous first-order optimality conditions of the continuous optimization problem (KKT conditions, Pontryagin maximum principle, ...) with state and control as unknowns,
- (d) or we write the continuous first-order optimality conditions of the continuous optimization problem with only the control as unknown.
- (c_1, c_2) We discretize the continuous first-order optimality conditions and solve the resulting extremal system according to a well chosen discretization space (e.g., shooting methods for optimal control in finite dimension).
- (c_3) When the Banach space U is Hilbert, we can identify its duality pairing with some inner product such that the derivative of the cost function can be identified with a gradient thanks to the Riesz representation theorem.
- (c_4) We then compute discretized derivatives according to a well chosen discretization and give them as inputs to `IpOpt` in `FreeFEM`.
- (e) When we compute the gradient of a function by semi-automatic differentiation, we can in some cases exhibit an adjoint variable which verifies the adjoint equation coming from the first-order optimality conditions. For instance in Appendix A.3, we use the idea of automatic differentiation in reverse mode to compute by hand the gradient of the cost function. We

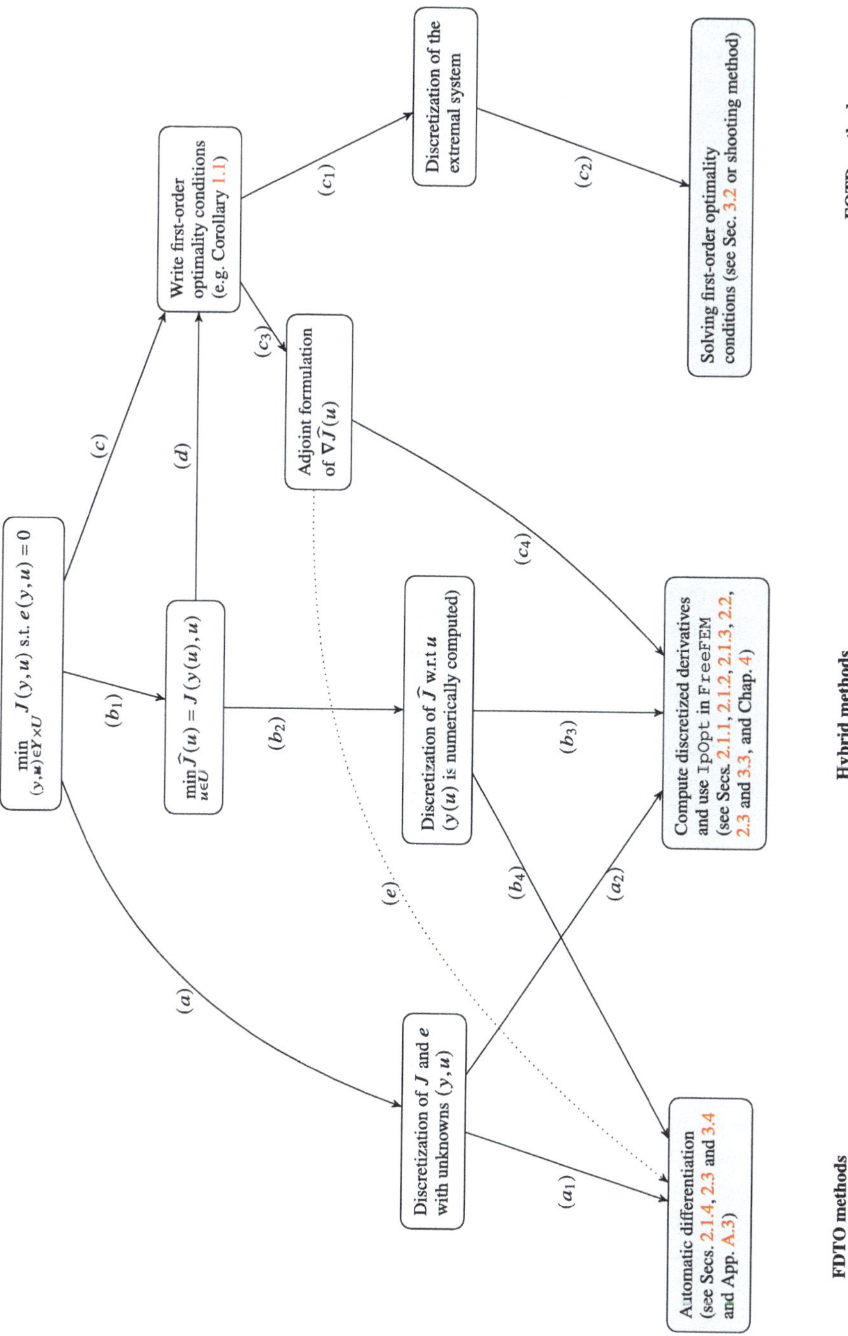

Fig. 1.2 Summary diagram of the differentiable optimization methods

then show that the gradient involves an adjoint variable which satisfies a discretized version of the continuous adjoint equation. This means that, with an appropriate discretization, automatic differentiation and hybrid methods lead to the same discretized gradient.

The Optimization Routine `IpOpt`

`IpOpt` (for **I**nterior **P**oint **Opt**imizer) is an open source software package for large-scale differentiable optimization problems. First developed in Fortran, the `C++` version allows to `IpOpt` to be more easily interfaced with `Matlab`, `Python`, `R`, `Julia`, etc.

`IpOpt` is included (with most of available linear solvers and options) in `FreeFEM`. We just have to call the library by adding in the head of the code the command: `load "ff-Ipopt"`. We previously highlighted the ability of `FreeFEM` to construct and manage sparse matrices. This is crucial for the efficiency of the optimization processing since `IpOpt` gathers many of the most powerful linear solvers (MUMPS, Pardiso, WSMP, HSL routines). A significant asset of `IpOpt` is its large panel of options (choice of linear solver, multipliers updating, adjustment of line search, BFGS or Newton method, etc). A full description of the interior point method can be found in [24] and options are available at [2, Ipopt Options Tab].

`IpOpt` is designed to find *local* solutions of mathematical optimization problems of the form

$$\min_{x\in\mathbb{R}^n} \quad f(x) \tag{1.24}$$

$$\text{s.t.} \begin{cases} g_L \leqslant g(x) \leqslant g_U \\ x_L \leqslant x \leqslant x_U \end{cases} \tag{1.25}$$

where $f : \mathbb{R}^n \to \mathbb{R}$ is the objective function and $g : \mathbb{R}^n \to \mathbb{R}^m$ stands for equality and inequality constraints. Here, g_L, g_U and x_L, x_U respectively refer to lower and upper bounds of constraints and variables. The functions f and g can be nonlinear and nonconvex, but are at least assumed to be twice differentiable. With this in mind, we aim to show how to transcribe a PDE optimization problem like (2.1, 2.2) to a finite-dimensional optimization problem like (1.24, 1.25). Once this transcription has been made, the most usual way to call `IpOpt` is:

```
IPOPT(f,df,d2f,C,jacC,x0,ub=xub,lb=xlb,cub=CUB,clb=CLB,optfile="
    ipopt.opt");
```

Code 1.5 Calling `IpOpt` in `FreeFEM` (see **ipoptfreefem.edp**) where `df` and `d2f` are respectively the gradient (an array) and the Hessian (a matrix in the Triplet Format, hopefully highly sparse) of the objective function; `C` includes the constraints g and `jacC` its Jacobian (a matrix in the Triplet Format, hopefully highly sparse too); `ub` and `lb` are respectively the upper and lower bounds for x, `cub` and `clb` are the upper and lower bounds of the constraints g, and `x0` is an initialization point; `optfile="ipopt.opt"` incorporates all options (maximum number of iteration, convergence tolerance threshold, choice of linear solver and so on). Difficulties for the computation of the Hessian `d2f` usually happen (too much memory greedy, needs the inverse of a matrix or slowing down too much the code), which is not disabling. `IpOpt` offers many options, we can bypass the problem by choosing a quasi-Newton method, which is the default option if `IpOpt` is called without specifying the Hessian.

Remark 1.8 Matrix triplet format is a way to numerically store sparse matrices, without specifying the zeros. Given a matrix, we only have to specify the row, the column, and the value for each non-zero element. This returns three arrays having the same size. `FreeFEM` will automatically deal with matrix triplet format through **`varf`** anf **`matrix`** commands.

```
IPOPT(f,df,C,jacC,x0,ub=xub,lb=xlb,cub=CUB,clb=CLB); // no
    Hessian ↦ BFGS
```

Despite its potential more expensive cost computation, the Newton method is usually converging to an optimal solution with less iterations than a quasi-Newton method. Nevertheless the latter sometimes offers more flexibility. We advise at least to specify the cost function `f`, the constraints `C` and their derivatives `df` and `jacC`, knowing that the more information we provide, the more efficient `IpOpt` is expected to be. We refer the reader to [24] to understand the particularities and how `IpOpt` works.

We mention two significant points: first, all `IpOpt` options are callable from `FreeFEM`. Second, the data to be given to `IpOpt` (`f`, `df`, `d2f`, `C`, `jacC`) need to respect a precise type: `df`,`C`,`x0`,`xub`,`xlb`,`CUB`,`CLB` have to be arrays and the matrices `d2f` and `jacC` have to be expressed in the Triplet Format for sparse matrices.

As an example, consider the minimization problem in $\mathbb{R}^2$

$$\min_{x\in K} f(x) = -x_1 x_2(1 - x_1 - x_2) \tag{1.26}$$

where

$$K = \left\{(x_1, x_2) \in \mathbb{R}^2 \mid x_1, x_2 \geqslant 0,\ x_1 + x_2 \leqslant 1\right\} \tag{1.27}$$

which is formulated in the template (1.24, 1.25) as

$$\min_{x_1,x_2} f(x) = -x_1x_2(1 - x_1 - x_2)$$

$$\begin{cases} g_L \leqslant g(x) = x_1 + x_2 \leqslant g_U \\ x_L \leqslant x \leqslant x_U \end{cases}$$

with $x_L = \begin{pmatrix} 0 \\ 0 \end{pmatrix}$, x_U large enough, $g_U = 1$ and g_L small enough. Derivatives are

$$\nabla f(x) = \begin{pmatrix} -x_2(1 - 2x_1 - x_2) \\ -x_1(1 - x_1 - 2x_2) \end{pmatrix}, \qquad \nabla g(x) = \begin{pmatrix} 1 \\ 1 \end{pmatrix},$$

$$\nabla^2 f(x) = -\begin{pmatrix} -2x_2 & 1 - 2x_1 - 2x_2 \\ 1 - 2x_1 - 2x_2 & -2x_1 \end{pmatrix}.$$

Therefore, the cost function is:

```
func real f(real[int] &X) //returns a real
{
    return -X[0]*X[1]*(1-X[0]-X[1]);
}
```

and its gradient and Hessian are:

```
func real[int] df(real[int] &X) // returns an array
{
    real[int] dJ(X.n); // size of X
    dJ = -1*[ X[1]*(1-2*X[0]-X[1]), X[0]*(1-2*X[1]-X[0])];
    return dJ;
}
matrix hess; // matrix has to be declared outside
func matrix d2f(real[int] &X) // returns a matrix
{
    hess = -1*[ [ -2*X[1] , 1-2*X[0]-2*X[1] ],
                [ 1-2*X[0]-2*X[1] , -2*X[0] ] ];
    return hess;
}
```

On the other side, the constraint function g is:

```
func real[int] C(real[int] &X) // returns an array
{
    real[int] cont(1); // array of size 1
    cont[0] = X[0]+X[1];
    return cont;
}
```

If only `X[0]+X[1]` is returned instead of the 1-size array containing this value, `IpOpt` will return an error. Like the Hessian of f, the Jacobian of g is:

```
matrix dc; // to be declared outside
func matrix jacC(real[int] &X) // returns a matrix
{
    dc = [[1,1]]; // double array of size (1,2)
    return dc;
}
```

We finally have to declare lower and upper bounds, an initialization point and then call `IpOpt`:

```
//Initialisation points
real[int] start = [1,1];
real[int] lm = [1]; // Lagrange mulitplier of constraint C
real[int] lz = [1,1]; // Lagrange mulitplier of simple lower
    bound constraint
real[int] uz = [1,1]; // Lagrange mulitplier of simple upper
    bound constraint

//Variables bounds
real[int] Xub = [10000,10000]; // has to be array
real[int] Xlb = [0,0]; // has to be array

//Constraints box1nds
real[int] Cub = [0]; // has to be array
real[int] Clb = [-10000]; // has to be array

//Calling IpOpt
IPOPT(J,gradJ,hessianJ,C,jacC,start,ub=Xub,lb=Xlb,clb=Clb,cub=Cub
    ,lz=lz,uz=uz,lm=lm,optfile="ipopt.opt");

cout << "(x,y)␣=␣" << "(" << start[0] << "," << start[1] << ")"
    << endl; // print
    solution
cout << "f(x,y)␣=␣" << J(start) << endl; // print solution
cout << "(lc)␣=␣" << lm << endl; // print constraint dual
    variable
cout << "(lz)␣=␣" << lz << endl; // print simple lower constraint
    dual variable
cout << "(uz)␣=␣" << uz << endl; // print simple upper constraint
    dual variable
```

which returns the optimal solution $x = (0.33, 0.33)$ in less than 10 iterations. We plot convergence curves in Fig. 1.3 for a desired convergence tolerance equal to 10^{-15}. The output of `IpOpt` is included in the `start` variables and returns the optimal solution if it is found. We also get the Lagrange multipliers through the variables `lm,lz,uz` highlighted above.

This example can be used as a template for other problems by adapting the formulation inside the functions (`f`, `df`, `d2f`, `C`, `jacC`). If `IpOpt` meets any difficulty to converge, or if it stops to a locally infeasible point, adjusting the starting

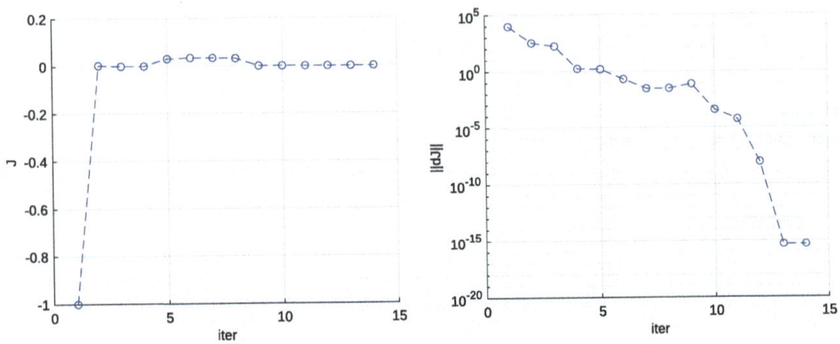

Fig. 1.3 Convergence curves: (**a**) objective function; (**b**) convergence criterion

point may sometimes help. In some cases, difficulties may occur when `IpOpt` requires to compute function derivatives, for instance for nonlinear PDEs or when state constraints are additionally given. An alternative is to compute derivatives by automatic differentiation. We may also implement finite differences but this requires a accurate approximation that may induce a crippling computation time.

Automatic Differentiation

We present here a brief introduction to automatic differentiation in direct and inverse modes. We refer the reader to [17] to go further and understand in depth the fundamentals of automatic differentiation.

The leading principle of automatic differentiation is that a function $f : \mathbb{R}^p \to \mathbb{R}^q$ given by a sequence of elementary numerical functio ns whose derivatives are already known can be differentiated by differentiating each line of the code step by step. This allows the chain rule to be used directly in the code and thus to compute the true derivative of the numerical function. A standard framework consists in considering a function $f : \mathbb{R}^n \to \mathbb{R}$ as mentioned in [5, 8] and a succession of calculation steps

$$(\phi_k, I_k)_{k\in\{0,\ldots,N\}} \in \mathcal{E}_{k-1}$$

with $\mathcal{E}_{k-1} = C \cup \mathcal{U}_k \cup \mathcal{B}_k$ where

$$C = \mathbb{R} \times \{\emptyset\}, \quad \mathcal{U}_k = U \times \{-n, \ldots, k\}, \quad \mathcal{B}_k = B \times \{-n, \ldots, k\}^2.$$

Here, U is a set whose elements are functions already known and implemented in the language and whose derivatives are still in the set U. Each step of the computation introduces a new variable and calls a known function of U according to a previous variable or a binary operation between two previous variables. Most of the time, this only allows classical analytic functions that we know how to differentiate analytically and whose derivatives are also combinations of elements of U (cos and sin for example). The set B gathers most of the binary operations $\{+, -, \times, \div\} \subset B$. We now suppose that we can find a finite sequence of operations

$$(\phi_k, I_k)_{k \in \{0,\dots,N\}}$$

such that all intermediate iterates given by

$$\forall k \in \{0, \dots, N\}, \quad x_k = \phi_k(x_{-n}, \dots, x_{-1}, \dots, x_{k-1})$$

lead to the final iterate

$$x_N = f(x_{-n}, \dots, x_{-1}).$$

In this framework $(x_{-n}, \dots, x_{-1})$ design inputs and $(x_k)_{k \geqslant 0}$ internal states in the coded function. The automatic differentiation uses the chain rule and the knowledge of the derivatives of the functions involved in ϕ_k to compute the final gradient. There are two methods. Either the chain rule is applied according to the calls of functions in series line by line from the first line, it is the direct method, or from the last line, it is the reverse method. The first one calculates the Jacobian of the function (so a matrix) while the last one gives the Jacobian applied in a given direction (so an array). The direct method gives directly the Jacobian matrix but is expensive because it is necessary to calculate all the intermediate derivatives which are not necessarily needed for the final direction.

Direct Mode

We give in the following lines a way to compute the function

$$f(u, v) = \Big((u + v)^2 + 1 + u\cos(u)\sin(v)\Big)^2$$

and its derivatives using automatic differentiation in direct mode. To do so, let us clarify some notations. The function f has two inputs (u, v) that corresponds in previous section to (x_{-2}, x_{-1}). Besides, f will ask in its coding way three computation steps. This is going to produce three variables (x, y, z) corresponding in the previous section to (x_0, x_1, x_2).

```
def f(u,v):
    x = u*cos(u)*sin(v)
    y = (u+v)**2 + 1
    z = (x+y)**2
    return z
```

Code 1.6 A way of coding function f

```
def df(u,v,du,dv):
    dx = u*sin(u)*du*sin(v)+cos(u)*du*sin(v)
         + u*sin(u)*cos(v)*dv
    x = u*cos(u)*sin(v)
    dy = 2*du*(u+v) + 2*dv*(u+v)
    y = (u+v)**2 + 1
    dz = 2*(x+y)*dx + 2*(x+y)*dy
    return dz
```

Code 1.7 Automatic differentiation in direct mode

Code 1.7 exhibits the functioning of an automatic differentiation tool in direct mode. Each calculation step is derivated with respect to the variables involved. For example the line `y = (u+v)**2 + 1` gives the line `dy = 2*du*(u+v)+ 2*dv*(u+v)`. At the end, we obtain a function that depends on (u, v) and on the several directions of descent (du, dv). To recover the Jacobian, we must then evaluate the function as many times as there are directions of descent, i.e. here $(1, 0)$ and $(0, 1)$. Direct mode is only based on the chain rules. The size of the matrix is not prohibitive in this case but we can also consider a function $f : \mathbb{R}^p \to \mathbb{R}^q$ with p, q large (this happens when discretizing the weak form of a PDE). We then have to compute a matrix of size $p \times q$ and evaluate it p times. In that case, direct mode is disabling. Fortunately, we do not usually need the full Jacobian, but only the Jacobian applied in a given direction. The reverse mode allows to compute the gradient without having to evaluate the Jacobian in each free direction. A parallel with the backpropagation method in machine learning is given in [6].

Reverse Mode

In the reverse mode case, we assume that a given function to be differentiated has been written in the following framework. We consider an evaluation sequence of intermediate operations $(\phi_k, I_k)_{k\in\{0..N\}}$ such that the function's output is

$$x_N = f(x_{-n}, \ldots, x_{-1}).$$

To compute the derivative of f in reverse mode, we associate to variables $(x_i)_{i\in\{-n,\ldots,N\}}$ a variable $(\lambda_i)_{i\in\{-n,\ldots,N\}}$, which we update with the following rule

$$\lambda_N = 1, \quad \forall i \in \{-n, \ldots, N-1\} \quad \lambda_i = \sum_{k>i} \lambda_k \partial_i \phi_k. \tag{1.28}$$

Given the previous rule, the partial derivatives of f evaluated at $(x_{-n}, \ldots, x_{-1})$ are stored in the variables $(\lambda_{-n}, \ldots, \lambda_{-1})$. The successive quantities λ_i thus depend on all $\lambda_{k\in\{i+1,\ldots,N\}}$ insofar as ϕ_k implies variables $x_{j\in\{i+1,\ldots,N\}}$. In practice, if the numerical code is not too long, we can proceed manually and line by line starting from the end of the implementation of the cost function. We then derivate

$$x_k = \phi_k(x_{-n}, \ldots, x_{k-1})$$

with respect to the involved variables $x_{-n}, \ldots, x_{k-1}$ and update the λ variables following the rule

$$\lambda_i \mathrel{+}= \lambda_k \partial_i \phi_k(x_{-n}, \ldots, x_{k-1}), \tag{1.29}$$

where $x+ = a$ means $x = x+a$. We highlight this method on the previous example. We must first rewrite f with a sequence of elementary operations and then apply (1.28). However, a more usual way of doing this is to differentiate each line of the code, from the last line to the first line, with respect to the variables involved and then update the corresponding adjoint variables.

```
def df(u,v):
    x = u*cos(u)*sin(v) # adjoint l_x
    y = (u+v)**2 + 1 # adjoint l_y
    cost = (x+y)**2 # adjoint l_c

    lc=1, lx=0, ly=0, lu=0, lv=0 # initialization of each λ
        for all variables involved
    # We now use (1.29) starting from the end of Code 1.6
    # Code 1.6 - line 4 involves (x, y)
    lx += 2*(x+y)*lc # line 4: ∂_x derivative, we update λ_x
    ly += 2*(x+y)*lc # line 4: ∂_y
    # Code 1.6 - line 3 involves (u, v)
    lu += 2*(u+v)*ly # line 3: ∂_u
    lv += 2*(u+v)*ly # line 3: ∂_v
    # Code 1.6 - line 2 involves (u, v)
    lu +=cos(u)*sin(v)*lx-u*sin(u)*sin(v)*lx # line 2: ∂_u
    lv += u*cos(u)*cos(v)*lx # line 2: ∂_v
    return lu,lv
```

Code 1.8 Automatic differentiation in reverse mode

The reader can see that all operations $x_k = \phi_k(x_{-n}, \ldots, x_{k-1})$ must be computed before the calculation of the quantities λ_i because they are required to update them. The price to pay is to write the code with a well-defined sequence of operations (ϕ_k, I_k) and to save all the intermediate variables $(x_1, \ldots, x_{N-1})$, which can be memory heavy if the function f to be calculated uses a large number of intermediate variables. It is therefore advisable to have a code that is "optimized". The generalization to a function with value in $(\mathbb{R}^q, q > 1)$ is done by permuting the quantities λ with q vectors fixed on the canonical basis.

Let us comment on the direct and reverse modes. The reverse mode is well suited when the function $f : \mathbb{R}^p \to \mathbb{R}^q$ to differentiate takes values in $\mathbb{R}^q$ with q small and especially when $p \gg q$. On the contrary, the direct mode is preferred when we differentiate with respect to few variables ($\mathbb{R}^p$ with p small and moreover when $p \ll q$). When considering optimal control problems, the choice of discretization can be different for functions and derivatives (indeed, an implicit scheme for the state usually implies an explicit scheme for the adjoint). The automatic differentiation therefore dispenses with this choice because both the direct and reverse methods return the true numerical derivative of the implemented function, and is therefore in line with a purely numerical approach.

A last important point that we would like to address is the close relationship between the inverse mode and the introduction of the adjoint variable in optimal control (see [20]). Let us consider the problem of controlled predator-prey equations (see [21, Ex. 4.10]), given by

$$\min\ J_T(u, v) = \frac{1}{2}\int_0^T (x(t) - 1)^2\, dt, \tag{1.30}$$

$$\text{subject to} \begin{cases} \dot{x} = x + y + u & x(0) = 1 \\ \dot{y} = x - y + v & y(0) = 1, \end{cases} \tag{1.31}$$

$$u, v \in \mathcal{U}_{ad} = \Big\{ f \in L^\infty(0, T) \mid \forall t \in (0, T)\ \ -1 \leqslant f(t) \leqslant 1 \Big\}, \tag{1.32}$$

we compute the derivative of the numerical implementation of the functional cost (the inverse mode is well suited since $J : \mathbb{R}^p \to \mathbb{R}$ with p large enough) and show that there appears a discretization of the adjoint equation coming from the Pontryagin maximum principle in finite dimension. The gradient computed by means of automatic differentiation gives a discretized adjoint equation whose implementation is described in Appendix A.2. Automatic differentiation also allows the computation of the Hessian, but the resulting matrix is usually not sparse and the optimization process thus becomes less efficient. Furthermore, automatic differentiation is difficult to implement with respect to mesh variations (hence, for general optimal shape design problems), for which the adjoint method is more appropriate.

Finally, in this chapter, we have provided most of the tools necessary to write a general PDE optimization problem in the form (1.7), to discretize it with well-chosen finite elements, and to numerically find an optimal solution using the interior point method `IpOpt`. In the following section, by taking several examples, we will give various ways to discretize an optimization problem governed by a partial differential equation.

References

1. FreeFem++: https://doc.freefem.org/documentation/index.html.
2. IpOpt : https://coin-or.github.io/Ipopt/
3. Fatiha Alabau-Boussouira, Yannick Privat, and Emmanuel Trélat. 2017. Nonlinear damped partial differential equations and their uniform discretizations. *Journal of Functional Analysis* 273(1): 352–403.
4. Thomas Apel, Mariano Mateos, Johannes Pfefferer, and Arnd Rösch. 2015. On the regularity of the solutions of Dirichlet optimal control problems in polygonal domains. *SIAM Journal on Control and Optimization* 53(6): 3620–3641.
5. Sylvain Auliac. 2014. *Développement d'outils d'optimisation pour freefem++*. Theses, Université Pierre et Marie Curie - Paris VI.
6. Atılım Günes Baydin, Barak A. Pearlmutter, Alexey Andreyevich Radul, and Jeffrey Mark Siskind. 2017. Automatic differentiation in machine learning: A survey. *Journal of Machine Learning Research* 18(1): 5595–5637.
7. Lawrence Craig Evans. 2010. *Partial differential equations*, volume 19 of Graduate studies in mathematics, 2nd edn. Providence: American Mathematical Society.
8. Andreas Griewank and Andrea Walther. 2008. *Evaluating derivatives*, 2nd edn. Philadelphia: Society for Industrial and Applied Mathematics (SIAM). Principles and techniques of algorithmic differentiation.
9. Nikolaus Hansen. 2016. The CMA evolution strategy: A tutorial. CoRR, abs/1604.00772.
10. Frédéric Hecht. 2012. New development in freefem++. *Journal of Numerical Mathematics* 20(3–4): 251–265.
11. Michael Hinze, Rene Pinnau, Michael Ulbrich, and Stefan Ulbrich. 2009. *Optimization with PDE constraints*, volume 23 of Mathematical modelling: Theory and applications. New York: Springer.
12. Robin Hogan. 2014. Fast reverse-mode automatic differentiation using expression templates in $C++$. *ACM Transactions on Mathematical Software* 40(4): Art. 26, 16.
13. Irena Lasiecka and Roberto Triggiani. 2000. *Control theory for partial differential equations: Continuous and approximation theories. I*, volume 74 of Encyclopedia of mathematics and its applications. Cambridge: Cambridge University Press. Abstract parabolic systems.
14. Irena Lasiecka and Roberto Triggiani. 2000. *Control theory for partial differential equations: Continuous and approximation theories. II*, volume 75 of Encyclopedia of mathematics and its applications. Cambridge: Cambridge University Press. Abstract hyperbolic-like systems over a finite time horizon.
15. Jacques-Louis Lions. 1971. *Optimal control of systems governed by partial differential equations*. Translated from the French by S. K. Mitter. Die Grundlehren der mathematischen Wissenschaften, Band 170. New York/Berlin: Springer.
16. Andrea Manzoni, Alfio Quarteroni, and Sandro Salsa. 2021. *Optimal control of partial differential equations: Analysis, approximation, and applications*. Springer: Berlin.
17. Richard D. Neidinger. 2010. Introduction to automatic differentiation and MATLAB object-oriented programming. *SIAM Review* 52(3): 545–563.
18. Pierre-Arnaud Raviart and Jean-Marie Thomas. 19863. *Introduction à l'analyse numérique des équations aux dérivées partielles*. Collection Mathématiques Appliquées pour la Maîtrise. [Collection of Applied Mathematics for the Master's Degree]. Paris: Masson.
19. Dominik Meidner Roland Becker and Boris Vexler. 2007. Efficient numerical solution of parabolic optimization problems by finite element methods. *Optimization Methods and Software* 22(5): 813–833.

20. Jesús María Sanz-Serna. 2016. Symplectic Runge-Kutta schemes for adjoint equations, automatic differentiation, optimal control, and more. *SIAM Review* 58(1): 3–33.
21. Emmanuel Trélat. 2005. *Contrôle optimal*. Mathématiques Concrètes. Paris: Vuibert. Théorie & applications.
22. Fredi Tröltzsch. 2010. *Optimal control of partial differential equations*, volume 112 of Graduate studies in mathematics. Providence: American Mathematical Society. Theory, methods and applications, Translated from the 2005 German original by Jürgen Sprekels
23. Marius Tucsnak and George Weiss. 2009. *Observation and control for operator semigroups*. Birkhäuser Advanced Texts: Basler Lehrbücher. Basel: Birkhäuser Verlag.
24. Andreas Wächter and Lorenz Theodor Biegler. 2006. On the implementation of a primal-dual interior point filter line search algorithm for large-scale nonlinear programming. *Mathematical Programming* 106(1): 25–57.

Chapter 2
PDE Constrained Optimization with `FreeFEM`

Abstract In this chapter, we consider several more or less classical PDE-constrained optimization problems, written in the framework introduced in Chap. 1, and we focus on their numerical solution using finite elements in `FreeFEM` and the method of interior points `IpOpt` respectively introduced in Sects. 1.2 and 1.2. Starting with a classical linear quadratic example, we give various ways for solving it, which can serve as templates for the user. We also explain how to use automatic differentiation in `FreeFEM`. Nonlinear and time-dependent PDEs are also considered to show the great efficiency of `FreeFEM` to handle the constraints induced by such PDEs. A brief subsection is devoted to showing how to solve optimal shape design problems within the optimal control viewpoint. We finally propose several numerical codes which are available at `FreeFEM`'s website, https://freefem.org/Optim/.

2.1 Linear Quadratic PDE Constrained Optimization

The first problem we address to illustrate the numerical solution of constrained PDE optimization is the minimization of a quadratic criterion subject to a linear elliptic equation. Some additional constraints are encoded in a convex set $\mathcal{U}_{ad}$. The weak formulation of the PDE is introduced in order to have a well adapted framework for its numerical solution as well as the writing of the derivatives and to design an appropriate numerical strategy.

Let Ω be a bounded Lipschitz domain and y_d be $L^2(\Omega)$. Let a be $L^\infty(\Omega)$ and positive, u_a, u_b be measurable functions and α a positive parameter. We seek an optimal solution $u \in L^2(\Omega)$ of

$$\min J(y,u) = \frac{1}{2}\int_\Omega (y(x) - y_d(x))^2\,dx + \frac{\alpha}{2}\int_\Omega u(x)^2\,dx \tag{2.1}$$

$$\text{subject to } \begin{cases} -\nabla\cdot(a\nabla y) = u & \text{in } \Omega, \\ y = 0 & \text{on } \partial\Omega. \end{cases} \tag{2.2}$$

$$\text{and } u_a \leqslant u \leqslant u_b. \tag{2.3}$$

F. Hecht et al., *PDE-constrained optimization within FreeFEM*, SpringerBriefs on PDEs and Data Science, https://doi.org/10.1007/978-981-95-7049-2_2

To formulate (2.1, 2.2) in the form of (1.7), the set of admissible controls is defined by

$$\mathcal{U}_{ad} = \{u \in L^2(\Omega) \mid u_a \leqslant u \leqslant u_b\} \subset U = L^2(\Omega).$$

The variational formulation of (2.2) consists in finding $y \in H_0^1(\Omega)$ solution of

$$\int_\Omega a\nabla y \cdot \nabla v \, dx - \int_\Omega uv \, dx = 0 \qquad \forall v \in H_0^1(\Omega). \tag{2.4}$$

We take $Y = H_0^1(\Omega)$. The Lax-Milgram Theorem (see [15, Lemma 1.8]) implies that, for any $u \in \mathcal{U}_{ad}$, there is a unique solution $y \in H_0^1(\Omega)$ of (2.4). Since $U = L^2(\Omega)$, the duality pairing is

$$\langle \cdot, \cdot \rangle_{U',U} = (\cdot, \cdot)_U .$$

Defining the operators

$$A \in \mathcal{L}\left(H_0^1(\Omega), H^{-1}(\Omega)\right) \quad \text{s.t.} \quad Ay : v \in H_0^1(\Omega) \mapsto \int_\Omega a\nabla y \cdot \nabla v \, dx,$$

$$B \in \mathcal{L}\left(L^2(\Omega)\right) \quad \text{s.t.} \quad Bu : v \in L^2(\Omega) \mapsto \int_\Omega uv \, dx,$$

the set Z must be defined so that the mapping

$$e : (y, u) \in H_0^1(\Omega) \times L^2(\Omega) \mapsto Ay - Bu \in Z$$

satisfies Assumption 1.3. The Gelfand triple

$$H_0^1(\Omega) \hookrightarrow L^2(\Omega) = L^2(\Omega)' \hookrightarrow H^{-1}(\Omega)$$

leads to set $Z = H^{-1}(\Omega)$ and the duality pairings $\langle \cdot, \cdot \rangle_{Y',Y}$ and $\langle \cdot, \cdot \rangle_{Z',Z}$ are thus compatible with the $L^2(\Omega)$-inner product. The last items of Assumption 1.3 follow from the Lax-Milgram Theorem. We have $A^* = A$ and $B^* = B$ and the adjoint p evolves in $Z' = H_0^1(\Omega)$. The partial derivatives of the functions under consideration are

$$\begin{aligned} \partial_y J(y, u) &= (y - y_d, \cdot)_U \\ \partial_u J(y, u) &= (\alpha u, \cdot)_U \\ \partial_y e(y, u) &= A \\ \partial_u e(y, u) &= -B. \end{aligned} \tag{2.5}$$

Now, from the numerical point of view, we have two main options: either the state equation $e(y, u) = 0$ is seen as a constraint to be checked and is considered, like the objective function J, to depend on both the state and the control (y, u). The optimization is then performed with respect to two optimization variables (state and control), this is Option 1; or else, the only optimization variable is the control, and in this case $e(y, u) = 0$ is preliminarily solved to compute $y(u)$ as a function of u, in order to express the reduced cost function $\widehat{J}(u)$, this is Option 2. Options 1 and 2 are also respectively called *simultaneous* and *sequential* methods.

Option 1: Variables $(y, u) \in Y \times U$

Cost: $J(y, u)$

Constraints: $e(y, u) = 0$ and $(y, u) \in Y_{ad} \times \mathcal{U}_{ad}$

Option 2: Variables $u \in U$

Cost: $J(y(u), u) = \widehat{J}(u)$

Constraints: $u \in \mathcal{U}_{ad}$

Option 2 brings up the reduced cost function $\widehat{J}$ whose derivatives with respect to the control variable u are computed using the adjoint representation presented in Sect. 1.2. The PDE constraint is thus implicitly contained in the numerical implementation of the reduced cost function. In contrast, Option 1 keeps the cost function dependent on the state and control variables (y, u) and the PDE constraint is an explicit equality constraint. Although Option 1 may appear more memory demanding, its efficiency compared to Option 2 depends on the specific problem and on the implementation details, for example in how the Hessian of the cost function is represented as a sparse matrix. A notable advantage of Option 1 is its ability to handle potential constraints on the state included in $\mathcal{Y}_{ad}$ whereas the adjoint equation and Option 2 are not well suited in this case. This being said, for the numerical part, let T_h be a triangulation of Ω:

```
mesh Th = square(50,50); //we take Ω =]0, 1[^2,
```

and the finite element space

$$V_h = \left\{ v \in H^1(\Omega), \quad \forall K \in T_h \;\; v_{|K} \in \mathbb{P}_1 \right\} = \text{Vect}(\phi_i)_{i \in \{1..n_d\}},$$

with $\mathbb{P}_1$ elements that guarantee the resulting linear system of (2.2) to be invertible (see [22]).

```
fespace Vh(Th,P1); // with P1 Lagrange finite elements
int nd = Vh.ndof;            // n_d degrees of freedom for Vh
```

Remark 2.1 Reminding the constraint (2.3), from the numerical point of view we usually compare the degrees of freedom instead of the functions themselves. This is exact for $\mathbb{P}_1$ elements and acceptable for most of Lagrange types elements. To this aim, we introduce the interpolation operator Π_h, which returns the approximated function on a given finite element space. The operator Π_h is related to the basis $(\phi_i)_{1\leqslant i\leqslant n_d}$, so that $\Pi_h(f) = \sum_{i=1}^{n_d} \phi_i'(f)\phi_i$. We obtain $(\phi_i')_{1\leqslant i\leqslant n_d}$ as a concatenation of the basis of the dual space of the chosen finite dimensional polynomial space, on each element of the triangulation, so that $\phi_i'(\phi_j) = \delta_{ij}$. Thus, to compare the two functions u and u_a we now have to compare their interpolates, respectively $u^h = \Pi_h(u)$ and $u_a^h = \Pi_h(u_a)$, or the degrees of freedom of the interpolates. We thus introduce the new operator $\mathcal{I}_h(f_h) = (\phi_i'(f))_i$ that returns the array of degrees of freedom of the finite element function f_h in the basis $(\phi_i)_{1\leqslant i\leqslant n_d}$. To simplify notations, we assume in the following that comparing two interpolates $u^h \leqslant v^h$ is done by comparing their degrees of freedom $\mathcal{I}_h(u^h) \leqslant \mathcal{I}_h(v^h)$. More generally, u^h will denote either the finite element function or its degrees of freedom $\mathcal{I}_h(u^h)$ according to the situation.

```
Vh uh; // u^h is a finite element function
real[int] Xu = uh[]; // Xu is the degrees of freedom of u_h
```

In the example above, we illustrate the difference between the finite element function u^h and its degrees of freedom. Throughout the book, since IpOpt only accepts arrays as inputs, we will deal with the degrees of freedom of the involved finite element functions.

Let us denote Y_h, U_h and the discrete $\mathcal{U}_{ad}^h$:

$$Y_h = \{v^h \in V_h, v^h_{|\partial\Omega} = 0\} = \mathrm{Vect}(\phi_i)_{1\leqslant i\leqslant n_d, i\notin\partial\Omega}$$

$$\mathcal{U}_{ad}^h = \{u^h \in V_h, u_a^h \leqslant u^h \leqslant u_b^h\} \subset U_h = V_h.$$

and the stiffness and mass matrices coming out from the operators A and B read

$$A_{h,ij} = \left(a\nabla\phi_i, \nabla\phi_j\right)_{(i,j)\in\{1..n_d\}^2}, \quad M_{h,ij} = \left(\phi_i, \phi_j\right)_{(i,j)\in\{1..n_d\}^2}.$$

One can notice that usually the size M_h is larger than the size A_h because the latter does not take into account the finite element functions which have non-zero values on the boundary. Fortunately, the way FreeFEM handles Dirichlet boundary conditions allows us to keep $Y_h = V_h$ and specify the Dirichlet boundary condition directly in the variational formulation in the Code 2.1. Which numerically results in specifying the Dirichlet condition value directly to the degrees of freedom whose finite elements is nonzero on the boundary. This adds in the matrix A_h some penalty terms at the indices related to the boundary elements (see FreeFEM's website for a detailed explanation on how FreeFEM manages Dirichlet boundary conditions).

```
Vh Y,V;
varf stiffness(Y,V) = int2d(Th)(grad(Y)'*grad(V))
                    - on(1,2,3,4,Y=0) //Homogeneous Dirichlet
                      condition
varf mass(Y,V) = int2d(Th)(Y*V);
matrix Mh = mass(Vh,Vh); // n_d × n_d matrix
matrix Ah = stiffness(Vh,Vh);
```

Code 2.1 Finite element matrices involved in (2.1,2.2) (see **lq_stationary_O1.edp** and **lq_stationary_O2.edp**)

`IpOpt` needs to compute the numerical derivatives of the cost and constraint functions. Having in mind Sect. 1.2, either the continuous derivatives of the functions are discretized according to a well-chosen scheme, or the functions are discretized first and their derivatives are computed later. In the first case, the advantage is to keep the structure of the continuous problem as long as possible. In the second case, we are able to return the true numerical derivatives.

Derivatives of Discretized Functions (FDTO)

In the *First Discretize Then Optimize* (*FDTO*) approach, the problem (2.1, 2.2) is discretized and yields a usual finite dimensional linear quadratic optimization problem. The mesh, finite element space and matrices were presented in the previous section. Depending on whether Option 1 or 2 is chosen, the discretized problem is

Option 1 :

$$\min_{(y_h,u_h)\in\mathbb{R}^{2n_d}} \frac{1}{2}(y_h - y_d^h)^T M_h(y_h - y_d^h) + \frac{\alpha}{2}u_h^T M_h u_h \tag{2.6}$$

$$\text{s.t.} \begin{cases} A_h y_h - M_h u_h = 0 \\ u_a^h \leqslant u_h \leqslant u_b^h, \end{cases} \tag{2.7}$$

Option 2 :

$$\min_{u_h\in\mathbb{R}^{n_d}} \frac{1}{2}(A_h^{-1}M_h u_h - y_d^h)^T M_h(A_h^{-1}M_h u_h - y_d^h) + \frac{\alpha}{2}u_h^T M_h u_h \tag{2.8}$$

$$\text{s.t. } u_a^h \leqslant u_h \leqslant u_b^h. \tag{2.9}$$

The matrices A_h and M_h have been defined in Code 2.1. They depend on the triangulation T_h. Whatever the choice of optimization variables, the numerical problem is written as

$$\min_X \ J(X) \quad \text{s.t.} \begin{cases} C_{lb} \leqslant C(X) \leqslant C_{ub} \\ X_{lb} \leqslant X \leqslant X_{ub}. \end{cases}$$

Option 1 : $X = (y_h, u_h) \in \mathbb{R}^{2n_d}$

$$J_h(X) = \frac{1}{2}(y_h - y_d^h)^T M_h(y_h - y_d^h) + \frac{\alpha}{2} u_h^T M_h u_h$$

$$C_h(X) = A_h y_h - M_h u_h$$

$$\nabla J_h(X) = \big(M_h(y_h - y_d^h)\ \alpha M_h u_h\big)$$

$$\nabla C_h(X) = \begin{pmatrix} A_h & 0_h \\ 0_h & M_h \end{pmatrix}$$

$$\nabla^2 J_h(X) = \begin{pmatrix} M_h & 0_h \\ 0_h & \alpha M_h \end{pmatrix}$$

Option 2 : $X = u_h \in \mathbb{R}^{n_d}$ (no explicit PDE constraints)

$$J_h(X) = \frac{1}{2}(A_h^{-1} M_h u_h - y_d^h)^T M_h (A_h^{-1} M_h u_h - y_d^h) + \frac{\alpha}{2} u_h^T M_h u_h$$

$$\nabla J_h(X) = M_h A_h^{-1} M_h (A_h^{-1} M_h u_h - y_d^h) + \alpha M_h u_h$$

$$\nabla^2 J_h(X) = M_h A_h^{-1} M_h A_h^{-1} M_h + \alpha M_h.$$

The functions required by IpOpt are expressed above in a matrix format. They must be written in FreeFEM following the model given in Sect. 1.2. In the following sections, we give details on how to implement these functions in FreeFEM according to the choice of discretizations.

Experience shows that Option 1 is, in this case, almost always the best solution. Since PDE constrained optimization uses many variables, the treatment of sparse matrices is a crucial point and is usually the best option in terms of computational speed and memory allocation. Although the number of optimization variables is larger than in Option 2, in Option 1 we do not have to compute the inverse of the matrix A_h, which would be a heavy task in high dimension. The matrices A_h and M_h are sparse while their inverse is not in general. Although the Hessian is constant in both situations, it is more memory greedy in the second case and involves more computation. Finally, for more complicated equations (such as nonlinear PDEs or time-dependent problems), it is not trivial that treating both state and control variables simultaneously is more efficient, except when state constraints are added to the problem. The adjoint method fits precisely in an approach where the previous one would not have worked. Getting some distance, it is not easy to know what options to choose. Given a boundary control problem for instance, we have much less unknows to find when we deal with the reduced cost function.

Discretization of Continuous Derivatives (FOTD)

Unlike the previous strategy, in the *First Optimize Then Discretize* (*FOTD*) approach, the continuous derivatives of the functions are first computed and then discretized. In this section, the problem (2.1, 2.2) is still linear quadratic. Note that when state constraints are added, the adjoint equation may become much more complicated. We focus on the option where only the control is the optimization variable and we write the continuous problem as

$$\min_{u \in \mathcal{U}_{ad}} J(y(u), u) = \widehat{J}(u).$$

As explained in Sect. 1.2, the adjoint approach facilitates the computation of the derivatives of the reduced cost function. The Lagrangian is

$$\begin{aligned} L : (y, u, p) \in Y \times U \times Z' \mapsto & \frac{1}{2}(B(y - y_d), y - y_d)_U \\ & + \frac{\alpha}{2}(Bu, u)_U + (p, Ay - Bu)_U, \end{aligned}$$

and we follow the steps (S.1, S.2, S.3, S.4). Let $u \in \mathcal{U}_{ad}$ be an optimal solution.

- S.1 The partial derivatives of J and e have been computed in (2.5). The operators A and B are selfadjoint ($A^* = A$ and $B^* = B$). Corollary 1.1 implies that if (y, u) is solution of (2.1, 2.2) then there exists an adjoint $p \in Z' = H_0^1(\Omega)$ such that

$$\begin{aligned} Ay &= Bu, \\ Ap &= -B(y - y_d), \\ (\alpha Bu - Bp, u)_U &\leqslant (\alpha Bu - Bp, v)_U \qquad \forall v \in \mathcal{U}_{ad} \end{aligned}$$

  which can be rewritten in the continuous form for $y \in H_0^1(\Omega)$ and $p \in H_0^1(\Omega)$ as

$$\begin{aligned} -\nabla \cdot (a \nabla y) = u \quad & \text{in } \Omega \\ y = 0 \quad & \text{on } \partial\Omega, \end{aligned}$$

$$\nabla \cdot (a \nabla p) = y - y_d \quad \text{in } \Omega$$
$$p = 0 \quad \text{on } \partial\Omega,$$
$$\int_\Omega (\alpha u - p) u \, dx \leqslant \int_\Omega (\alpha u - p) v \, dx \quad \forall v \in \mathcal{U}_{ad}.$$

- S.2 We seek the solution $y \in H_0^1(\Omega)$ of the state equation $e(y, u) = 0$

$$- \nabla \cdot (a \nabla y) = u \qquad \text{in } \Omega$$
$$y = 0 \qquad \text{on } \partial\Omega,$$

```
macro state() {
    solve State(Y,V)  = int2d(Th)(a*(grad(Y)'*grad(V)))
                      - int2d(Th)(U*V)
                      + on(1,2,3,4,Y=0); } //
```

Code 2.2 LQ state equation

which is equivalent to

$$y_h = A_h^{-1} M_h u_h.$$

- S.3 We seek the solution $p \in H_0^1(\Omega)$ of the adjoint equation

$$\nabla \cdot (a \nabla p) = y - y_d \quad \text{in } \Omega$$
$$p = 0 \quad \text{on } \partial\Omega,$$

```
macro adjoint() {
    solve Adjoint(P,Q)= int2d(Th)(a*(grad(P)'*grad(Q)))
                      + int2d(Th)((Y-Yd)*Q)
                      + on(1,2,3,4,P=0); } //
```

Code 2.3 LQ adjoint equation

which is equivalent to

$$p_h = -A_h^{-1} M_h (y_h - y_d^h).$$

- S.4 The first derivative of the reduced cost function is given by

$$\left(\nabla \widehat{J}(u), q\right)_U = \int_\Omega (\alpha u - p) q \, dx,$$

```
// matrix for the numerical gradient (for L²(Ω) product, the
    matrix is the mass matrix)
varf vB(Y,V) = int2d(Th)(Y*V);
matrix M = vB(Vh,Vh,solver=sparsesolver);
macro interpgrad() {
    solve L2grad(theta,V) = int2d(Th)(theta*V) - int2d(Th)((alpha
        *U-P)*V);
    real[int] dJ = M*theta[]; } //
```

Code 2.4 LQ gradient's interpolation

which finally returns $\alpha u - p$ for the $L^2(\Omega)$-inner product, i.e. $M_h(\alpha u_h - p_h)$. If we consider another cross-product, it would necessary to adapt the resulting matrix. The previous steps S.2, S.3, S.4 performed successively return the same gradient as in Option 2 of the previous *FDTO* approach by performing the same operations. This illustrates the fact that *FODT* and *FDTO, Option2* can lead to the expression of the same gradient. This can be numerically verified specifying the option `derivative_test first-order` in the "ipopt.opt" file.

Let us give a brief explanation of the numerical step S.4. The real number $\langle D\widehat{J}(u), q\rangle_{U',U}$ is expressed using the duality pairing $\langle \cdot, \cdot \rangle_{U',U}$ which here corresponds to the $L^2(\Omega)$-inner product and thus brings out the gradient $\nabla\widehat{J}(u)$. As seen in Sect. 1.2, the routine **varf** gives the matrix of a given variational formulation. Hence, writing

```
varf derive(V,Q)= int2d(Th)((alpha*U-P)*Q)
```

is numerically similar to recovering the continuous linear form

$$q \mapsto \int_\Omega (\alpha u - p) q \, dx$$

in a well-chosen finite element space basis. Thus `dJ` denotes the interpolation of this linear form in this basis of the finite element space. One could of course choose another inner product, whose matrix in the finite element space is P_h, which would not return

$$\alpha u_h - p_h$$

but rather

$$P_h^{-1} M_h(\alpha u_h - p_h).$$

This is a crucial point to understand because many problems show the derivative of the reduced cost function in a linear form which must then be interpolated thanks to a good discretization of the space U. Usually, the additional constraints included in $\mathcal{U}_{ad}$ are specified in another function C, the so-called constraint function (e.g., $u \mapsto C(u) = \int_\Omega u(x)\, dx$). The Jacobian of C is then required by `IpOpt`. The only remaining task is to express the cost function and its gradient following the steps (S.2, S.3, S.4) on the triangulation T_h respectively in the Codes 2.5 and 2.6.

```
func real J(real[int] &X)
{
    U[] = X; // associated array X to the control u function.
    state; // returns y solution of (2.2)
    return int2d(Th)((Y-Yd)^2) + alpha*int2d(Th)(U^2);
}
```

Code 2.5 LQ cost function (see **lq_stationary_indirect.edp**)

Having in mind Rem. 2.1, recall that IpOpt can only take arrays as inputs. Consequently, the cost function J takes as input the array of the degrees of freedom of the finite element function U. Then, we compute inside J the cost function with the finite element function U. Hence the cost function depends on an array and can be called from IpOpt.

```
func real[int] dJ(real[int] &X)
{
    U[] = X;
    state;     // state equation (see Code 2.2)
    adjoint; // adjoint equation (see Code 2.3)
    interpgrad; // interpolation of the gradient (see Code 2.4)
    return dJ;
}
```

Code 2.6 Derivative of the LQ cost function (see **lq_stationary_indirect.edp**)

In the first method, the computation of the Hessian is rather easy but it requires a matrix inversion. We advise instead to compute the theoretical first-order derivatives and to use the BFGS approximation of the Hessian provided by IpOpt.

Remark 2.2 The optimality conditions produce a state-adjoint system called *extremal system*. In finite-dimensional optimal control, the solution of the first-order optimality system can be computed by implementing an indirect shooting method (see [23, Chap. 9]) in which, if for example the initial and final states are fixed, one has to properly adjust the initial adjoint vector so that, when integrating the extremal with the corresponding initial state and adjoint, the final state matches the desired value. Numerically, the shooting method consists in combining a differential equation integrator with a Newton method. When the optimal control problem is solved in finite dimension and this dimension is not too large, the method is feasible and, when properly initialized to ensure convergence, it provides a fast and accurate solution. But in high dimension, it can be extremely difficult to initialize the method successfully. This is particularly true in the case of PDE optimization where the size of the adjoint variable is related to the size of the mesh. Therefore, most of the time, it is not realistic to ensure the convergence of a shooting method for optimal PDE control problems.

Inhomogeneous Dirichlet Boundary Conditions

The previous example treated the case of homogeneous Dirichlet conditions, which makes easier the numerical study because the variational formulation can be written on well-identified functional spaces (this is also the case when dealing with Neumann or Robin's boundary conditions). We now generalize the above numerical approaches to inhomogeneous Dirichlet conditions. Instead of (2.2), we now consider

$$\begin{cases} -\nabla \cdot (a \nabla y) = u & \text{in } \Omega \\ y = g & \text{on } \partial\Omega, \end{cases} \tag{2.10}$$

where g is assumed to be smooth enough. The way to numerically manage the boundary condition depends on the choice of either Option 1 or Option 2.

In Option 2, the boundary condition can be immediately reported in the variational formulation so that the numerical state equation becomes:

```
macro state() {
    solve State(Y,V) = int2d(Th)(a*(grad(Y)'*grad(V)))
                     - int2d(Th)(U*V)
                     + on(1,2,3,4,Y=g); //y = g on ∂Ω
} //
```

Code 2.7 State equation

The adjoint equation remains unchanged but the adjoint is well modified because it still depends on the state.

In contrast, Option 1 requires more work. For homogeneous Dirichlet conditions, the stiffness matrix A_h introduced in Code 2.1 forces y to be equal zero on the boundary $\partial\Omega$ and can only be applied to finite element functions in $H_0^1(\Omega)$. Since y now belongs to an affine subset of $H^1(\Omega)$ and not to $H_0^1(\Omega)$, the state equation cannot be written as before. The variational formulation of (2.10) is now:

$$\text{find } y \in H^1(\Omega), \int_\Omega a \nabla y \cdot \nabla v \, dx - \int_\Omega uv \, dx = 0 \qquad \forall v \in H_0^1(\Omega). \tag{2.11}$$

A numerical trick consists in introducing, using the Dirichlet map $\mathbf{D}$ (see [25, Sec. 10.6]), the solution $\mathbf{D}g$ of

$$-\Delta(\mathbf{D}g) = 0 \text{ on } \Omega \text{ and } \mathbf{D}g = g \text{ on } \partial\Omega$$

and its numerical version:

```
Vh Dg;
solve dirmap(u,v) = int2d(Th)(grad(Y)'*grad(V)) + on(1,2,3,4,Y=g)
    ;
Dg[] = Y[];
```

Code 2.8 Dirichlet map

We seek the solution of (2.10) in the affine space $H_0^1(\Omega) + \mathbf{D}g$ under the form $y = z + \mathbf{D}g$. Therefore, finding a solution $y \in H^1(\Omega)$ of (2.10) is equivalent to finding a solution $z \in H_0^1(\Omega)$ of (2.2). The optimization variables are (z, u) and the discretized problem is

$$\min_{(z_h,u_h)\in\mathbb{R}^{2n_d}} J_h(z_h, u_h) = \frac{1}{2}\left(z_h + (\mathbf{D}g)_h - y_d^h\right)^T M_h \left(z_h + (\mathbf{D}g)_h - y_d^h\right) + \frac{\alpha}{2} u_h^T M_h u_h$$

$$\text{s.t.} \begin{cases} A_h z_h - M_h u_h = 0 \\ u_a^h \leqslant u_h \leqslant u_b^h. \end{cases}$$

The state constraint is unchanged (as well as its Jacobian) while the cost function now involves $y_h = z_h + (\mathbf{D}g)_h$. Most of the following examples specify homogeneous Dirichlet conditions but a generalization to inhomogeneous Dirichlet conditions can be made along the above lines.

Automatic Differentiation Alternative

Unless the derivatives are easy to calculate, it may be advantageous to use automatic differentiation, which allows to compute the numerical derivatives at the computer accuracy. As automatic differentiation is not available in FreeFEM (especially when one wants to make a derivation with respect to the mesh points), it is required to export the problem data collected with FreeFEM (mesh data, matrices of the variational forms involved, etc.) to another language which benefits from an automatic differentiation program.

In order to keep a user-friendly interface, we propose two practical solutions: either to use the modeling language AMPL, or the Python package CasADi (with GNU license and callable from Matlab or Python). AMPL is a commercial software but is free of access on the NEOS server. Finally, for those who are quite familiar with the language C++, we advise to combine directly IpOpt with an automatic differentiation tool C++ (such as CppAD or Adept, see [16]) for more efficiency.

In the following, we provide numerical examples that illustrate how to combine `FreeFEM` with `AMPL` in the linear quadratic case (2.1, 2.2). The combination with the Python package `CasADi` is presented in App. A.3.

For a given triangulation T_h and generated with `FreeFEM`, we store the matrices A_h and M_h constructed in the Code 2.1 of Sect. 2.1 in sparse matrices via the COO (Coordinate) format. The files `"A.txt"` and `"B.txt"` are generated via `FreeFEM` with the command (see **lq_stationary_AMPL.edp**):

```
{ofstream fout("Ah.txt");
fout << Ah << endl;
}
```

`AMPL` ("A Mathematical Programming Language", see [1, 13]) is a highly developed software for modeling and solving large-scale optimization problems. Like `CasADi`, the unknown variables must be declared and the objective and constraints must be defined as a nonlinear programming problem. The transcription of `AMPL` is based on classical logical operators, aggregation functions and sets, while the transcription of `CasADi` is done more in the framework of matrix calculus. Both handle the sparsity feature of matrices very well. As far as `AMPL` is concerned, it is necessary to write the problem using appropriate sets (set of indices of the non-zero elements of a matrix for example). A significant advantage of `AMPL` is the possibility to call several recognized optimization solvers, like `Knitro`, `CPLEX`, `IpOpt`, etc, which can be used free of charge on the server NEOS. The program is usually divided into three files: "file.mod" contains the model, while "file.dat" gathers the parameter allocations and "file.run" the successive commands.

The Code 2.9 includes the introduction of variables (lines 12 and 13), parameters (line 1 and lines 6 to 10), minimization of the objective function (line 15) and constraint functions (lines 19 to 21). Unlike `CasADi`, we do not use explicit matrix calculus in `AMPL`, the **sum** operator on appropriate sets is preferred instead. The parameters `A` and `M` represent the sparse matrices A_h and M_h introduced in the Code 2.1. To preserve sparsity, the matrices in `AMPL` are downloaded from a text file in triplet format (see sparse matrices on the `IpOpt` web page for multiple ways to store them). The indices for which `A` has non-zero values are declared in a two-dimensional set `indexA`. The parameter `A{indexA}` depending on this set thus gathers the non-zero values of the matrix A_h (idem for M_h). The quadratic cost function is obtained on line 15 of the Code 2.9 by summing over all the indices `(i,j) in indexM` and we therefore do not take into account in the sum the indices where `M` has a null value. This induces a significant gain in computation time.

```
1 param m integer >=0;
2 set index = 0..m;
3 set indexA dimen 2;
4 set indexM dimen 2;
5
```

```
param alpha =0.1;
param A{indexA}; # stiffness matrix
param M{indexM}; # mass matrix
param L{index};
param yd; # target

var Y{index}; # state
var U{index}; # control

minimize quad: sum{(i,j) in indexM}(
    0.5*(Y[i]-yd)*M[i,j]*(Y[j]-yd) +
    0.5*alpha*(U[i]*M[i,j]*U[j]));

subject to PDE{i in index}: sum{(i,j) in
    indexA}(A[i,j]*Y[j]) - sum{(i,j) in indexM}(M[i,j]*U[j])
    = 0;

subject to lowbound{i in index}: U[i] >= 0;
subject to upbound{i in index}: U[i] <=1;

subject to volume: sum{i in index} U[i]*L[i] == 0.25; #
    ∫_Ω u(x) dx = 0.25
```

Code 2.9 AMPL: "file.mod" (see **ffad.{dat,mod,run}**)

Like all the parameters, the data are assigned in Code 2.10:

```
data;

param m:= 2600;
param: indexA: A:= include A.txt;

param: indexM: B:= include B.txt;

param yd:= 0.1;
read{i in index}(L[i]) < L.txt;
```

Code 2.10 AMPL: "file.dat" (see **ffad.{dat,mod,run}**)

AMPL is then called by typing in the terminal the command: `ampl file.run`

```
model file.mod;
data file.dat;

option solver ipopt;
option ipopt_options"max_iter=1000 tol=1.e-12
    linear_solver=mumps";

solve;
```

Code 2.11 AMPL: "file.run" (see **ffad.{dat,mod,run}**)

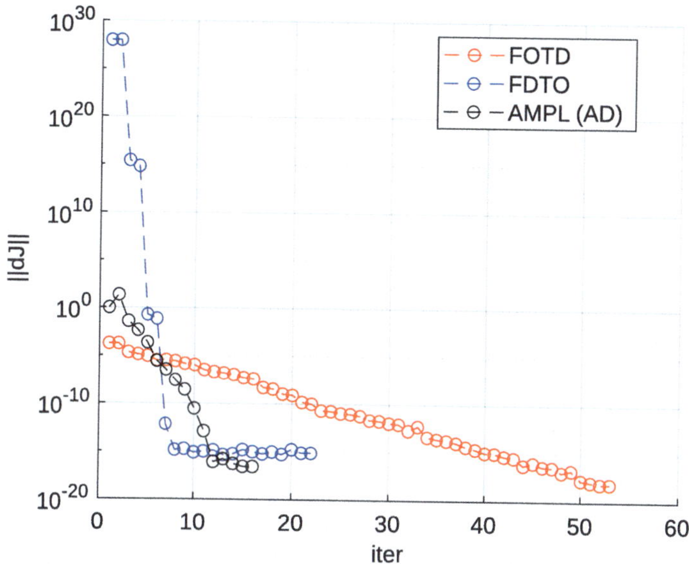

Fig. 2.1 Convergence curves of several methods

Provided that the data can be easily imported as in Code A.3, `AMPL` stands out for its ability to manage many efficient optimization solvers such as `Knitro`, `CPLEX` and so on, which, as already said, can be used for free on NEOS, on which we have to upload the three above files (in order to be uploaded to NEOS, data file must contain parameters allocations stated in file *"A.txt"*, *"B.txt"* etc, since they cannot be uploaded aside). `Python` (and `Matlab`) alternatives are presented in Appendix A.3.

To conclude this part, we plot in Fig. 2.1 the convergence curves of the scaled error respectively associated to the several methods illustrated above. In the case of the Linear-Quadratic problem (2.1, 2.2), we notice that the option 1 of the *FDTO* method is ten times faster than the *FOTD* one. This is explained through the fact that the option in *FDTO* method allows the use of the Hessian and thus Newton method instead the L-BFGS alternative endowed in `IpOpt`. `AMPL` endowed with its own automatic differentation tool is an interesting acceptable alternative.

2.2 Extension to Time-Dependent Problems

In this section, we show how to deal with a time-dependent optimal control problem, which consists in determining $u \in L^2((0,T); L^2(\Omega))$ solution of

$$\min J(y,u) = \frac{1}{2}\int_0^T \int_\Omega (y(x,t) - y_d(x))^2 \, dxdt + \frac{\alpha}{2}\int_0^T \int_\Omega u(x,t)^2 \, dxdt \tag{2.12}$$

$$\text{subject to } \begin{cases} y_t - \nabla \cdot (a\nabla y) = u & \text{in } (0,T) \times \Omega \\ y = 0 & \text{in } (0,T) \times \partial\Omega \\ y(0) = y^0 \\ u_a \leqslant u \leqslant u_b \end{cases} \tag{2.13}$$

where $y^0 \in L^2(\Omega)$ is fixed.

For the existence of solutions, we refer the reader to [15, 20].

We focus here on the numerical implementation for solving this problem. Several time discretization strategies can be implemented, which can then be combined with the optimization strategies described previously. We follow Option 1, but the other possibilities can be adapted in a similar way to time-dependent problems. We introduce a mesh of the domain Ω, as in the previous stationary example. The main issue is how to deal simultaneously with temporal and spatial discretizations.

A first classical time discretization consists in using an implicit Euler scheme combined (we will mostly choose implicit schemes since there is no CFL condition to be satisfied) with finite elements $\mathbb{P}_1$ in space.

A second possibility is to use the ability of the FreeFEM to handle 3D problems by considering a 3D time-space mesh and to discretize simultaneously time and space variables with finite elements $\mathbb{P}_1$.

Implicit Euler Scheme

Although an implicit scheme is a bit harder to implement numerically because it requires a matrix inversion, its advantage over explicit schemes is that it does not require any CFL condition. We consider the mesh $\mathcal{T}_h$ introduced in Sect. 2.1 and the matrices A_h and M_h. Given an integer n_t, consider a time subdivision

$$t_0 = 0 < t_1 < \cdots < t_{n_t} = T$$

of the interval $[0, T]$. We introduce the discrete variables

$$\tilde{Y} = (y_1, \cdots, y_{n_t}) \in \mathbb{R}^{n_d n_t} \text{ and } \tilde{U} = (u_1, \cdots, u_{n_t}) \in \mathbb{R}^{n_d n_t},$$

with $y_i, u_i \in \mathbb{R}^{n_d}$, arrays of degree of freedom of the finite element approximations of $y(t_i, \cdot)$ and $u(t_i, \cdot)$ respectively. We then discretize the problem (2.12, 2.13) as

$$\min\ J(\tilde{Y}, \tilde{U}) = \frac{1}{2}\sum_{k=1}^{n_t}(y_k - y_d^h)^T M_h (y_k - y_d^h) + \frac{\alpha}{2}\sum_{k=1}^{n_t} u_k^T M_h u_k \tag{2.14}$$

$$\text{subject to} \begin{cases} M_h \dfrac{y_{k+1} - y_k}{\delta t} + A_h y_{k+1} = M_h u_{k+1}, & \forall k \in (0 \ldots n_t - 1) \\ y_0 = y^0 \\ u_a^h \leqslant u_k \leqslant u_b^h, & \forall k \in \{0 \ldots n_t\}. \end{cases} \tag{2.15}$$

We simplify the notations by introducing the matrices

$$A_t = A_h + \frac{1}{\delta t} M_h, \quad M_t = \frac{1}{\delta t} M_h$$

for $\delta t = \frac{T}{n_t}$ and the sparse matrices

$$\tilde{A} = \underbrace{\begin{pmatrix} I_{n_d} & 0_{n_d} & \cdots & 0_{n_d} \\ -M_t & A_t & \cdots & 0_{n_d} \\ \vdots & \ddots & \ddots & \vdots \\ 0_{n_d} & \cdots & -M_t & A_t \end{pmatrix}}_{n_d n_t \text{ columns and rows}} \quad \tilde{M} = \begin{pmatrix} 0_{n_d} & \cdots & 0_{n_d} \\ M_t & \ddots & 0_{n_d} \\ \vdots & \ddots & \vdots \\ 0_{n_d} & \cdots & M_t \end{pmatrix}$$

$$\tilde{D} = \begin{pmatrix} M & 0_{n_d} & \cdots & 0_{n_d} \\ 0_{n_d} & M & \ddots & 0_{n_d} \\ \vdots & \ddots & \ddots & \vdots \\ 0_{n_d} & \cdots & 0_{n_d} & M \end{pmatrix} \tag{2.16}$$

to reformulate the optimization problem (2.14, 2.15) as the problem of determining $(\tilde{Y}, \tilde{U}) \in \mathbb{R}^{2n_d(n_t+1)}$ solution of

$$\min\ J(\tilde{Y}, \tilde{U}) = \frac{1}{2}(\tilde{Y} - Y_d)^T \tilde{D}(\tilde{Y} - Y_d) + \frac{\alpha}{2}\tilde{U}^T \tilde{D}\tilde{U}$$

$$\text{subject to} \begin{cases} \tilde{A}\tilde{Y} - \tilde{M}\tilde{U} = \begin{pmatrix} y^{0,h} \\ 0_{n_d(n_t-1)} \end{pmatrix}, \\ u_a^h \leqslant \tilde{U}_k \leqslant u_b^h \quad \forall k \in \{0..n_t\}. \end{cases}$$

We thus obtain a finite-dimensional constrained linear quadratic problem

$$\nabla J(\tilde{Y}, \tilde{U}) = \begin{pmatrix} \tilde{D}(\tilde{Y} - Y_d) \\ \alpha \tilde{D} \tilde{U} \end{pmatrix}, \quad \nabla^2 J(\tilde{Y}, \tilde{U}) = \begin{pmatrix} \tilde{D} & O \\ 0 & \alpha \tilde{D} \end{pmatrix},$$

$$\nabla C(\tilde{Y}, \tilde{U}) = \begin{pmatrix} \tilde{A} & 0 \\ 0 & -\tilde{M} \end{pmatrix}.$$

for which the numerical implementation in `FreeFEM` now follows the method presented in Sect. 2.1, based on the template introduced in Sect. 1.2 (see **lq_time_O1.edp** and **lq_time_O2.edp**).

Time Discretization with `FreeFEM`

Another possibility is to exploit the ability of `FreeFEM` to handle multidimensional problems, by considering the time variable as a third variable of the z space and transforming a 2D (or 1D) problem into a 3D (or 2D) problem. In the example (2.12, 2.13), the variational formulation reads as follows:

$$\int_{(0,T)\times\Omega} (y_t v + a \nabla y \cdot \nabla v) \, dxdt = \int_{(0,T)\times\Omega} uv \, dxdt \tag{2.17}$$

for all v in $H^1((0,T) \times \Omega))$ such that $v_{|(0,T)\times\partial\Omega} = 0$. We build a mesh of the 3D domain $(0, T) \times \Omega$ on which the PDE is settled. In order to manage the 3D meshes, we use at the beginning of the file the command: **`load`** *"msh3"*. Several options to build a 3D mesh with `FreeFEM` are available. We prefer to start from an initial mesh of the domain Ω, subset of$\mathbb{R}^2$, that we extend (along the axis z) to a mesh of the cylinder $(0, T) \times \Omega$ thanks to the command **`buildlayers`** (Fig. 2.2):

```
mesh Th2 = square(20,20);
int[int] rup=[0,30],rdown=[0,20],rmid=[1,10,2,10,3,10,4,10];
mesh3 Th = buildlayers(Th2,20,zbound=[0,T],labelmid=rmid,labelup=
    rup,labeldown=rdown);
```

Code 2.12 3D mesh cylinder with 2D mesh basis (see **lq_time_t_as_z.edp**)

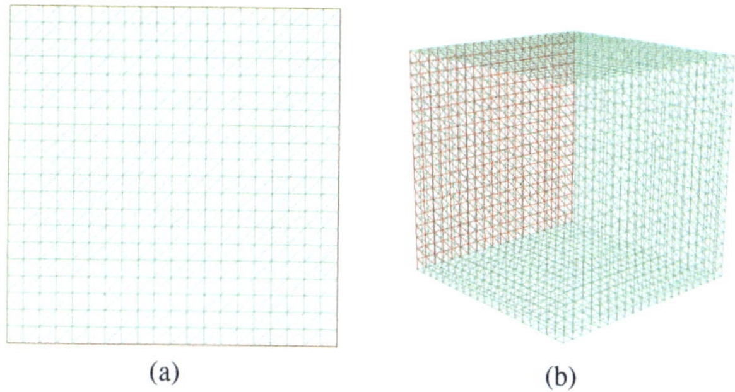

(a) (b)

Fig. 2.2 (**a**) initial square mesh; (**b**) 3D mesh with **buildlayers**

The new finite element space is

$$V_h = \left\{ v \in H^1((0,T)\times\Omega), \quad \forall K \in T_h \;\; v_{|K} \in \mathbb{P}_1 \right\}$$

As in the static case, we define the matrix of the variational formulation (2.17):

```
varf vA(Y,V) = int3d(Th)(-dz(V)*Y+a*(grad(Y)'*grad(V))) //
     dz(Y)=∂t y
              + on(10,Y=0) // Dirichlet boundary condition
              + on(20,Y=Y0); // initial condition y0
matrix Ah = vA(Vh,Vh,solver=sparsesolver);
```

We define the matrix of the cost function:

```
varf vCost(Y,V) = int3d(Th)(Y*V);
matrix Mh = vCost(Vh,Vh,solver=sparsesolver);
```

Given a mesh of the domain $(0,T)\times\Omega$, we now write the discretized optimal problem as

$$\min_{(\tilde{Y},\tilde{U})\in\mathbb{R}^{2N}} \frac{1}{2}(\tilde{Y}-Y_d)^T M_h(\tilde{Y}-Y_d) + \frac{\alpha}{2}\tilde{U}^T M_h \tilde{U}$$

$$\text{s.t.} \begin{cases} A_h\tilde{Y} = M_h\tilde{U} \\ U_a^h \leqslant \tilde{U} \leqslant U_b^h \end{cases}$$

where N is the number of degrees of freedom of the chosen finite element space (taking into account the time dimension), U_a^h and U_b^h are the resulting arrays of degrees of freedom from the interpolation operator connected to the finite element space $\mathbb{V}_h$ of the functions u_a and u_b respectively. Here, N is of the same order as the quantity $n_d(n_t+1)$ from the previous section. The gradient and Hessian of the cost function and the Jacobian of the constraint function are now easy to compute and the numerical implementation follows what has been presented in Sect. 2.1.

When considering linear quadratic optimal control problems, the problems resulting from the various discretization strategies remain general high-dimensional linear quadratic optimization problems. Nevertheless, the previous examples can be used as models for other examples, more complicated, but which require a significant upstream work on the chosen discretization.

When the PDE under consideration is linear, we recommend to use Option 1. Indeed, keeping the PDE as a linear constraint rather than solving it directly, which requires at least a matrix inversion (the matrix inversion will always take place but the linear solver chosen in `IpOpt` will do it), will most of the time bring two advantages. First, it is easier to compute the Jacobian of this constraint and it is almost always easier to compute the derivatives of the cost function with respect to state and control than the derivatives of the reduced cost function, whose Hessian is even harder to compute. Second, we keep the sparse feature of the matrices involved in the discrete writing of the problem.

Option 2 (use of an adjoint representation) is more suitable when the dependence on the control is more complex and cannot be expressed easily, as in the case of shape optimal design, when the control is a shape (see [3, 14]) or for nonlinear problems. Option 1 and Option 2 differ in the choice of the parameterization of the optimal shape to be found.

Furthermore, regardless of the chosen approach, time-dependent problems are much more computationally and memory intensive and require careful attention to potential simplifications.

2.3 Optimization Under Semilinear PDE Constraints

We add a nonlinear term to the previous problem. Given $y_d \in L^2(\Omega)$ and $u_a, u_b \in L^\infty(\Omega)$, we seek an optimal solution $u \in L^2(\Omega)$ of

$$\min\ J(y,u) = \frac{1}{2}\int_\Omega (y(x) - y_d(x))^2\, dx + \frac{\alpha}{2}\int_\Omega u(x)^2\, dx \tag{2.18}$$

$$\text{subject to} \begin{cases} -\nabla\cdot(a\nabla y) + \phi(y) = u & \text{in } \Omega \\ y = 0 & \text{on } \partial\Omega \\ u_a \leqslant u \leqslant u_b. \end{cases} \tag{2.19}$$

Following the framework stated in [24, Assumptions 4.2 and 4.3], the function $\phi : \mathbb{R} \to \mathbb{R}$ is continuous, monotone increasing, globally bounded and twice differentiable. The global boundedness assumption is only used to ensure that $\phi(y) \in L^2(\Omega)$. In our numerical tests, we choose $\phi(y) = y^3$, which is not globally bounded, but Sobolev embeddings guarantee that $H^1(\Omega) \hookrightarrow L^6(\Omega)$ for a bounded Lipschitz domain Ω of $\mathbb{R}^2$ or $\mathbb{R}^3$. Therefore, given any $u \in L^2(\Omega)$, the PDE (2.19) has a unique weak solution $y \in H_0^1(\Omega)$ (see [24, Remarks on Theorem 4.4]). Compared with the state (2.2), the semilinear PDE is more difficult to solve with the finite element method. We do not have any linear variational formulation and therefore there is no way to solve it with a single matrix inversion.

We propose to solve the PDE numerically with successive approximations (iterative fixed point) or with a Newton method and to use an adjoint representation to compute the derivatives. Note that the adjoint only requires a matrix inversion because the adjoint equation is linear. Let T_h be the same triangulation of the domain Ω as before and let V_h still denote the finite element space $\mathbb{P}_1$. The variational formulation is

$$\int_\Omega a\nabla y \cdot \nabla v\, dx + \int_\Omega \phi(y)v\, dx = \int_\Omega uv\, dx \qquad \forall v \in H_0^1(\Omega).$$

A first possibility is to use Algorithm 2 (fixed point) to compute a numerical solution of this variational formulation, whose translation in `FreeFEM` macros is given in Code 2.13.

Algorithm 2 Fixed point method for semilinear PDEs

set $err = 1$
while $(err > 10^{-10})$ **do**
  solve: $\forall v \in H_0^1(\Omega), \quad \int_\Omega a\nabla z \cdot \nabla v\, dx + \int_\Omega \phi(y^{p-1})v\, dx = \int_\Omega uv\, dx$
  compute $err = \|z - y^{p-1}\|_{L^2(\Omega)}$
  set $y^p = z$
  set $p = p + 1$.
end while

```
macro semilinearstateFP(){
  real err=1;
  while (err>tol) {
    Vh Yold;
    solve semilinear(Y,V) = int2d(Th)(a*grad(Y)'*grad(V))
                          + int2d(Th)(Yold^3*V)  // φ(y) = y^3
                          - int2d(Th)(U*V)
                          + on(1,2,3,4,Y=0);
    real[int] taberr = Y[]-Yold[];
    err = l2norm(taberr); // ||y^{k+1} - y^k||_{L^2(Ω)}
    // l2norm defined in appendix A.1
    }
} //
```

Code 2.13 Semilinear state equation with a fixed point method (see **semilinear_fixedpoint_ipopt.edp**)

Another possibility is to implement a Newton-Raphson strategy, consisting in determining $y \in V$ such that $F(y) = 0$ with $F : V \mapsto V$, as done in Algorithm 3.

Algorithm 3 Newton method

set $err = 1$
set y_0 and w initialized
while ($err > 10^{-10}$) **do**
  $y^p = y^{p-1} - w$
  solve: $DF(y^p)w = F(y^p)$
  $err = \|v\|$
  set $p = p + 1$.
end while

In the context of the problem (2.18, 2.19), F is the variational formulation of the state equation (2.19) and we have

$$F(y) = \int_\Omega (a\nabla y \cdot \nabla v + \phi(y)v - uv)\,dx$$

$$DF(y)w = \int_\Omega (a\nabla w \cdot \nabla v + \phi'(y)wv)\,dx$$

for some $v \in H_0^1(\Omega)$, and (2.19) is solved in Code 2.14.

```
macro semilinearstateNewton(){
  real err=1;
  Vh W=0;
  while (err>tol) {
    Y[] -= W[]; // y^{p+1} = y^p - w
    solve semilinear(W,V) =
          int2d(Th)(a*grad(W)'*grad(V)    // DF(y)w
             + 3*Y^2*W*V) // φ(y) = y^3
        - int2d(Th)(a*grad(Y)*grad(V) + Y^3*V - U*V) //F(y)
        + on(1,2,3,4,W=0);
    err = l2norm(W[]); // ||w||_{L^2(Ω)}
    // l2norm defined in appendix A.1
    }
} //
```

Code 2.14 Semilinear state equation with Newton method (see **semilinear_fixedpoint+Newton_ipopt.edp**)

One may wonder whether it is better to use a fixed point or a Newton method. Newton's algorithm is well known for its fast, but convergence can in general only be obtained if its initialization is close enough to the solution. The fixed point algorithm is less sensitive to the initialization constraint but the convergence is slower in general. One may consider a hybrid method consisting in using a fixed point method in the first iterations and then switching to a Newton method when one is close enough to the solution (this is the idea of global or damped Newton methods).

Remark 2.3 We note that, in the case of the nonlinear equation of state, solutions using the fixed-point or Newton methods are iterative methods that may require many iterations. The equation of state is sometimes solved several times (cost and gradient function), whereas it would be useful to keep the current state value so as not to have to solve the equation of state again when calculating the adjoint. An alternative is proposed in the Code **semilinear_fixedpoint_ipopt_with_less_computations.edp** to decrease the calculation time extended by the state equation being solved too many times. To do this, we added a condition whereby if the state is very close to the previously calculated state, the equation of state is not recalculated.

As in Sect. 2.1, Y is the set $H_0^1(\Omega)$ and the set of admissible controls $\mathcal{U}_{ad}$ is the subset of $L^2(\Omega)$ given by

$$\mathcal{U}_{ad} = \left\{u \in L^2(\Omega), u_a \leqslant u \leqslant u_b\right\} \subset U = L^2(\Omega),$$

$Z = H^{-1}(\Omega)$ (dual of $H_0^1(\Omega)$), so that Assumption 1.3 is satisfied, and

$$J(y,u) = \frac{1}{2}\int_\Omega (y(x) - y_d(x))^2\, dx + \frac{\alpha}{2}\int_\Omega u(x)^2\, dx,$$

$$e(y,u) = Ay + B\phi(y) - Bu.$$

Under [24, Assumptions 4.14], existence of an optimal control $\bar{u} \in \mathcal{U}_{ad}$ is guaranteed and the optimality conditions stated in Corollary 1.3 give

$$y \in H_0^1(\Omega) \text{ solution of: } -\nabla\cdot(a\nabla y) + \phi(y) = u \quad \text{in } \Omega, \tag{2.20}$$

$$p \in H_0^1(\Omega) \text{ solution of: } \nabla\cdot(a\nabla p) - \phi'(y)p = y - y_d \quad \text{in } \Omega, \tag{2.21}$$

$$u \in \mathcal{U}_{ad} \text{ such that: } (\alpha u - p, v - u)_U \geqslant 0 \quad \forall v \in \mathcal{U}_{ad}. \tag{2.22}$$

Again, like in the example (2.1, 2.2), the duality pairing $\langle\cdot,\cdot\rangle_{U',U}$ is compatible with the $L^2(\Omega)$-inner product and hence the adjoint representation yields the gradient of the reduced cost function

$$\left(\nabla\widehat{J}(u), v\right)_U = \int_\Omega (\alpha u - p)v\, dx \quad \forall v \in L^2(\Omega).$$

Although the state equation is nonlinear, the adjoint equation is linear and can thus easily be solved with FreeFEM:

```
macro semilinearadjoint() {
    solve SLAdjoint(P,Q)= int2d(Th)(a*(grad(P)'*grad(Q)))
                        + int2d(Th)(3*Y^2*P*Q)  // φ'(y) = 3y²
                        + int2d(Th)((Y-Yd)*Q)
                        + on(1,2,3,4,P=0);
} //
```

Code 2.15 Semilinear adjoint equation

At this step, it suffices to rewrite the cost function and its derivative as in Codes 2.5 and 2.6 by replacing the macros of the previous state and adjoint equations.

In the case of semilinear equations, it is possible to work with sparse matrices that do not depend on the state and control, provided that we are careful enough on how to discretize the nonlinear term. Therefore, we can override the computation of the adjoint by using automatic differentiation with the software AMPL as in Sect. 2.1. We thus reduce the cost function to the quadratic discretized cost

$$J_h(y_h, u_h) = \frac{1}{2}(y_h - y_d^h)^T M_h (y_h - y_d^h) + \frac{\alpha}{2} u_h^T M_h u_h$$

while the state equation is viewed as a constraint

$$e_h(y_h, u_h) = A_h y_h + M_h (y_h)^3 - M_h u_h$$

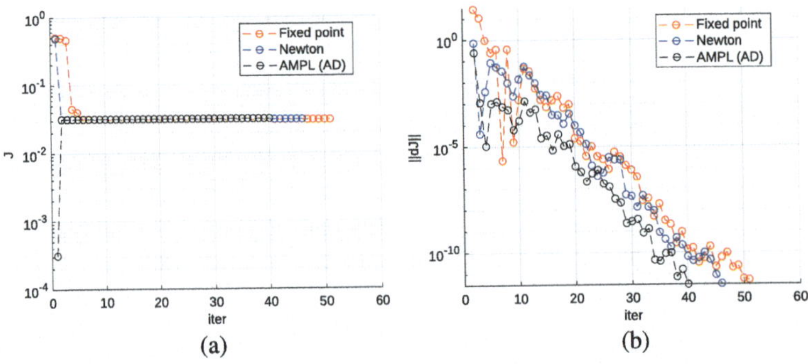

Fig. 2.3 Convergence curves for semilinear case: (**a**) objective function; (**b**) convergence criterion

so that "file.mod" becomes:

```
param m integer >=0;
set index = 0..m;
set indexA dimen 2;
set indexM dimen 2;

param alpha =0.1;
param A{indexA}; # stiffness matrix
param M{indexM}; # mass matrix
param L{index};
param yd; # target

var Y{index}; # state
var U{index}; # control

minimize quad: sum{(i,j) in indexM}(
      0.5*(Y[i]-yd)*M[i,j]*(Y[j]-yd) +
      0.5*alpha*(U[i]*M[i,j]*U[j]));

subject to PDE{i in index}: sum{(i,j) in indexA}(A[i,j]*Y[j])
                          + sum{(i,j) in indexM}(M[i,j]*Y[j]**3)
                          - sum{(i,j) in indexM}(M[i,j]*U[j]) =0;

subject to lowbound{i in index}: U[i] >= 0;
subject to upbound{i in index}: U[i] <=1;

subject to volume: sum{i in index} U[i]*L[i] == 0.25; #
      ∫_Ω u(x) dx = 0.25
```

Code 2.16 AMPL: "file.mod" - semilinear case

The convergence curves of the various methods are shown in Fig. 2.3. The methods converge more or less quickly to the same solution. Here we emphasize the advantages of automatic differentiation over adjoint methods that bypass solving the state equation by iteration methods. The convergence of the algorithm is then faster and its numerical writing is much easier. Nevertheless, we must pay some attention to the numerical discretization of the cost and state constraint functions. Indeed, we will show in Chap. 3 the importance of the choice of finite element spaces for the algorithm to give an acceptable solution.

2.4 Optimal Shape Design Problems

FreeFEM is adapted to solve numerically optimal shape design problems. Indeed, FreeFEM is user-friendly for solving PDEs, but also for building and modifying meshes. Shape optimization is a vast field and there are many ways to compute numerical optimal solutions. It is possible to implement geometric and topological optimization as well as homogenization methods. For the *level-set* method in

FreeFEM, we refer the reader to [4, 5] and to the dedicated website. We also mention [9] which is a pedagogical introduction to the level-set method in shape optimization using FreeFEM. In short words, level-set method is related to the solution of an advection equation combined with a gradient descent algorithm (see [11] for an example of an efficient gradient descent algorithm applied to shape optimization). We mention two methods that can be combined with IpOpt:

- Shape deformation methods: we look for a way to modify an initial shape to a so-called optimal shape and thus compute the derivatives of the cost and constraint functions using classical shape optimization analysis tools. The command movemesh provided by FreeFEM allows us to write an optimal shape design problem in the context of Hadamard boundary variations. The derivatives with respect to the domain and the appropriate deformation vector fields are thus computed to find the next iteration using the movemesh. We refer to [14] for an introduction to shape variation strategies.
- Relaxation methods: we look for an optimal solution in a larger admissible set containing the classical design sets and expect the solution to be in the starting set of shapes. One of the best known methods is homogenization (see [2, 6]). The optimal solution often presents a gray level instead of being black or white, which underlines the appearance of a relaxation phenomenon.

The example of Sect. 4.1 further is devoted to illustrating the first method. Here, let us quickly describe the second method, which can be implemented in the framework of the numerical methods of Sect. 2.1 by considering a variant of the linear quadratic example (2.1, 2.2). Let Ω be a subset of $\mathbb{R}^2$ and let $y_d \in L^2(\Omega)$ be a target function. The set of admissible controls is

$$\mathcal{U}_{ad} = \{\omega \subset \Omega \text{ measurable}\,, \quad |\omega| \leqslant \omega_0\}$$

where $|\omega|$ is the Lebesgue measure of ω. Denoting by χ_ω the indicator function (defined by $\chi_\omega(x) = 1$ if $x \in \omega$ and 0 otherwise), we seek for a measurable domain ω solution of

$$\min_{\omega \in \mathcal{U}_{ad}} J(y,\omega) = \frac{1}{2}\int_\Omega (y(x) - y_d(x))^2\, dx$$

$$\text{s.t.} \begin{cases} -\Delta y = \chi_\omega & \text{in } \Omega \\ y = 0 & \text{on } \partial\Omega. \end{cases}$$

We refer the reader to [18] for sufficient conditions ensuring existence of solutions. For the numerical solving, we relax the indicator function χ_ω of ω as shown in Fig. 2.4.

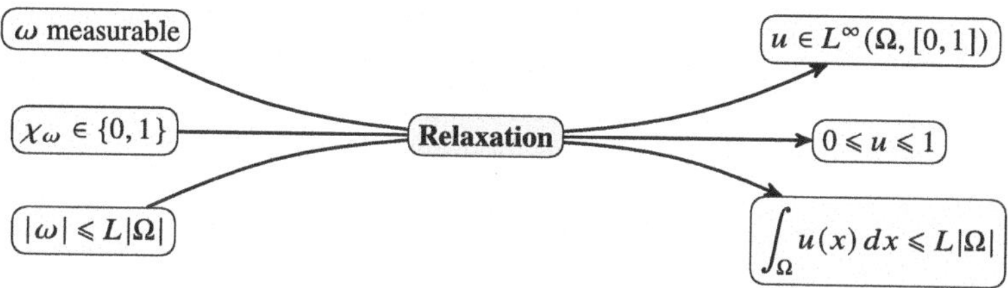

Fig. 2.4 Convexification strategy

This leads to the relaxed (convexified) problem

$$\min_{u \in \mathcal{U}_{ad}} J(y,u) = \frac{1}{2}\int_\Omega (y(x) - y_d(x))^2\,dx$$

$$\text{s.t.} \begin{cases} -\Delta y = u & \text{in } \Omega \\ y = 0 & \text{on } \partial\Omega \\ 0 \leqslant u \leqslant 1 \end{cases}$$

with

$$\mathcal{U}_{ad} = \left\{ u \in L^2(\Omega), \quad 0 \leqslant u \leqslant 1 \text{ and } \int_\Omega u(x)\,dx = L|\Omega| \right\}.$$

s We now adapt the example (2.1, 2.2). Shape optimization is performed by stipulating upper and lower bound constraints on the control u in the hope that they will be saturated for the optimal solution (if not, this means that there is relaxation). The generalization of such methods can be found in the *homogenization* method studied in [2]. The optimization solver `IpOpt` may be used.

2.5 Conclusion

In this chapter, we have considered several PDE-constrained optimization problems and solved them with `FreeFEM`. We have provided the detail of all steps, starting from an appropriate mathematical formulation of the problem to the numerical implementation. We have designed several methods depending on the chosen problem. We stress that `FreeFEM` has already been the focus of several educational papers, such as [17, 26] in topology optimization, [10] in geometric optimization, [7] for structural optimization and [12] where the authors combined `FreeFEM` with `IpOpt`.

`FreeFEM` is also powerful for three-dimensional problems. Let us mention [8] where the authors build finite element method for fast rotating Bose-Einstein condensates. Regarding optimization problems, we mention [19] where the authors propose a new framework for the two- and three-dimensional topological optimization of the weakly-coupled fluid–structure system. In the same field, we cite [21] where the authors present an educative paper and focus on a level-set method in dimension two with an extension to the three-dimensional case. In the sequel, we will focus on more specific problems, following the methods illustrated in this chapter.

References

1. AMPL: https://ampl.com/products/ampl/.
2. Grégoire Allaire. 2002. *Shape optimization by the homogenization method*, volume 146 of Applied mathematical sciences. New York: Springer.
3. Grégoire Allaire. 2007. *Conception optimale de structures*, volume 58 of Mathématiques & applications (Berlin) [Mathematics & applications]. Berlin: Springer. With the collaboration of Marc Schoenauer (INRIA) in the writing of Chapter 8.
4. Grégoire Allaire, Frédéric de Gournay, François Jouve, and Anca-Maria Toader. 2005. Structural optimization using topological and shape sensitivity via a level set method. *Control Cybernetics* 34(1): 59–80.
5. Grégoire Allaire, François Jouve, and Anca-Maria Toader. 2004. Structural optimization using sensitivity analysis and a level-set method. *Journal of Computational Physics* 194(1): 363–393.
6. Grégoire Allaire, Arnaud Münch, and Francisco Periago. 2010. Long time behavior of a two-phase optimal design for the heat equation. *SIAM Journal on Control and Optimization* 48(8): 5333–5356.
7. Grégoire Allaire and Olivier Pantz. 2006. Structural optimization with FreeFem++. *Structural and Multidisciplinary Optimization* 32(3): 173–181.
8. Ionut Danaila and Frédéric Hecht. 2010. A finite element method with mesh adaptivity for computing vortex states in fast-rotating Bose–Einstein condensates. *Journal of Computational Physics* 229(19): 6946–6960.
9. Charles Dapogny and Florian Feppon. 2023. Shape optimization using a level set based mesh evolution method: an overview and tutorial. *Comptes Rendus. Mathématique* 361: 1267–1332.
10. Charles Dapogny, Pascal Frey, Florian Omnès, and Yannick Privat. 2018. Geometrical shape optimization in fluid mechanics using freefem++. *Structural and Multidisciplinary Optimization* 58(6): 2761–2788.
11. Florian Feppon, Grégoire Allaire, and Charles Dapogny. 2020. Null space gradient flows for constrained optimization with applications to shape optimization. *ESAIM - Control, Optimisation and Calculus of Variations* 26: Paper No. 90, 45.
12. Francisco Javier Fernández, Lino José Alvarez-Vázquez, Áurea Martínez, and M.E. Vázquez-Méndez. 2017. A 3d optimal control problem related to the urban heat islands. *Journal of Mathematical Analysis and Applications* 446(2): 1571–1605.
13. Robert Fourer, David Gay, and Brian Kernighan. 1989. *Algorithms and model formulations in mathematical programming*. chapter AMPL: A mathematical programming language, 150–151. Berlin/Heidelberg: Springer.
14. Antoine Henrot and Michel Pierre. 2018. *Shape variation and optimization*, volume 28 of EMS tracts in mathematics. Zürich: European Mathematical Society (EMS).
15. Michael Hinze, Rene Pinnau, Michael Ulbrich, and Stefan Ulbrich. 2009. *Optimization with PDE constraints*, volume 23 of Mathematical modelling: Theory and applications. New York: Springer.
16. Robin Hogan. 2014. Fast reverse-mode automatic differentiation using expression templates in $C++$. *ACM Transactions on Mathematical Software* 40(4): Art. 26, 16.

17. Cheolwoong Kim, Mingook Jung, Takayuki Yamada, Shinji Nishiwaki, and Jeonghoon Yoo. 2020. Freefem++ code for reaction-diffusion equation–based topology optimization: for high-resolution boundary representation using adaptive mesh refinement. *Structural and Multidisciplinary Optimization* 62(1): 439–455.
18. Gontran Lance, Emmanuel Trélat, and Enrique Zuazua. 2020. Shape turnpike for linear parabolic PDE models. *IEEE Control Systems Letters* 142: 104733.
19. Hao Li, Tsuguo Kondoh, Pierre Jolivet, Kozo Furuta, Takayuki Yamada, Benliang Zhu, Kazuhiro Izui, and Shinji Nishiwaki. 2022. Three-dimensional topology optimization of a fluid–structure system using body-fitted mesh adaption based on the level-set method. *Applied Mathematical Modelling* 101: 276–308.
20. Jacques-Louis Lions. 1971. *Optimal control of systems governed by partial differential equations*. Translated from the French by S. K. Mitter. Die Grundlehren der mathematischen Wissenschaften, Band 170. New York/Berlin: Springer.
21. Jorge Morvan Marotte Luz Filho, Raquel Mattoso, and Lucas Fernandez. 2023. A freefem code for topological derivative-based structural optimization. *Structural and Multidisciplinary Optimization* 66(4): 74.
22. Pierre-Arnaud Raviart and Jean-Marie Thomas. 1983. *Introduction à l'analyse numérique des équations aux dérivées partielles*. Collection Mathématiques Appliquées pour la Maîtrise. [Collection of Applied Mathematics for the Master's Degree]. Paris: Masson.
23. Emmanuel Trélat. 2005. *Contrôle optimal*. Mathématiques Concrètes. Paris: Vuibert. Théorie & applications.
24. Fredi Tröltzsch. *Optimal control of partial differential equations*, volume 112 of Graduate studies in mathematics. Providence: American Mathematical Society. Theory, methods and applications, Translated from the 2005 German original by Jürgen Sprekels.
25. Marius Tucsnak and George Weiss. 2009. *Observation and control for operator semigroups*. Birkhäuser advanced texts: Basler Lehrbücher. Basel: Birkhäuser Verlag.
26. Benliang Zhu, Xianmin Zhang, Hai Li, Junwen Liang, Rixin Wang, Hao Li, and Shinji Nishiwaki. 2021. An 89-line code for geometrically nonlinear topology optimization written in freefem. *Structural and Multidisciplinary Optimization* 63(2): 1015–1027.

Chapter 3
Plateau Problem

Abstract In this chapter, we address a problem of calculus of variations: compute the optimal solution of the Plateau problem. We present three different methods illustrated in Fig. 1.2 of Chap. 1: a direct method via automatic differentiation, an indirect method by solving the optimality conditions and a hybrid method. We finally highlight the non-commutativity of the previous methods which do not necessarily yield the same solution, depending on the choice of discretization.

3.1 Introduction

The Plateau problem is named after the Belgian natural scientist J. Plateau, who made numerous experiments with soap films, realizing a large variety of minimal surfaces. We can find a complete description and many solving suggestions, e.g., in [1, 2]. We illustrate on such an example of variational computation how the combination of `FreeFEM` and `IpOpt` can improve the numerical solution for a PDe constrained optimization problem. Depending on a good choice of discretization, we will also show that automatic differentiation is a powerful tool.

In this well known problem, the aim is to find a minimal surface whose boundary is fixed. We restrict ourselves to closed and C^2 surfaces, parametrized by means of the graph of a given function $u : \bar{\Omega} \to \mathbb{R}$ (with $\Omega \subset \mathbb{R}^2$) belonging to the admissible set

$$S(\Omega) = \Big\{ u : \bar{\Omega} \to \mathbb{R}, \quad u_{|\Omega} \in C^2(\Omega, \mathbb{R}), \quad u_{|\partial\Omega} \in C^1(\partial\Omega, \mathbb{R}) \Big\},$$

so that an admissible surface is given by the parametric surface

$$X(x, y) = (x, y, u(x, y)), \quad (x, y) \in \bar{\Omega}.$$

The cost functional is

$$J(u) = \int_\Omega (1 + |\nabla u|^2)^{\frac{1}{2}} \, dx.$$

F. Hecht et al., *PDE-constrained optimization within FreeFEM*, SpringerBriefs on PDEs and Data Science, https://doi.org/10.1007/978-981-95-7049-2_3

Given a function $\gamma \in C^1(\partial\Omega, \mathbb{R})$ along the boundary of Ω, we seek an optimal solution of the Plateau optimization problem

$$\min_{u \in S(\Omega)} J(u) \text{ such that } u_{|\partial\Omega} = \gamma. \tag{3.1}$$

We propose three methods for its numerical solving:

- The first one consists in finding the critical points of the functional J thanks to an iterative procedure combining a fixed point method and a Newton method. This method is part of *FOTD*.
- The second one calls the `IpOpt` solver from `FreeFEM`.
- In the third one, we use automatic differentiation with `AMPL` after having generated in `FreeFEM` the discretization of the functional J with finite elements.

Note that `IpOpt` requires to compute at least the gradient, and if possible, the Hessian. To this aim, we compute the first derivative of the cost functional J

$$DJ(u)v = \int_\Omega \frac{\nabla u \cdot \nabla v}{(1 + |\nabla u|^2)^{\frac{1}{2}}}\, dx \quad \forall v \in S(\Omega) \tag{3.2}$$

and its second order derivative

$$D^2J(u)(v, w) = \int_\Omega \frac{\nabla v \cdot \nabla w}{(1 + |\nabla u|^2)^{\frac{1}{2}}} - \frac{(\nabla u \cdot \nabla v)(\nabla u \cdot \nabla w)}{(1 + |\nabla u|^2)^{\frac{3}{2}}}\, dx \quad \forall v, w \in S(\Omega). \tag{3.3}$$

We introduce a triangulation T_h of the domain Ω and will induct hereafter several finite element spaces. We define the `FreeFEM` macros of the cost function and its derivatives that we need for the various approaches. As an example, we define Ω as the unit square and

$$\gamma(x, y) = \cos(\pi x)\cos(\pi y).$$

We approximate the functions u in $S(\Omega)$ with $\mathbb{P}_1$ finite elements (we will choose as well $\mathbb{P}_2$ and will compare solutions):

```
mesh Th=square(10,10,[x+1,y]);
fespace Vh(Th,P1);

Vh u,up,v;
func gamma = cos(pi*x)*cos(pi*y);
macro grad(u) [dx(u),dy(u)] //
```

Code 3.1 Minimal surface numerical framework

Besides, we introduce the macros of the derivatives of the cost function J that we need for the optimization methods.

```
macro J(u)            int2d(Th)(sqrt(1.0 + grad(u)'*grad(u))) //
macro DJ(u,v)         int2d(Th)(grad(u)'*grad(v)/sqrt(1.0 + grad(u)
    '*grad(u))) //
macro DJfp(u,up,v)    int2d(Th)(grad(u)'*grad(v)/sqrt(1.0 + grad(up
    )'*grad(up))) // for fixed point method
macro D2J(u,v,w)      int2d(Th)(grad(w)'*grad(v)/sqrt(1.0 + grad(u)
    '*grad(u)) -  grad(u)'*grad(v)*grad(u)'*grad(w)/sqrt(1.0 +
    grad(u)'*grad(u))^3 ) //
```

Code 3.2 Minimal surface macros

3.2 Solving with an Iterative Method

We claim that (3.1) is a convex optimization problem so that solving the first-optimality conditions leads to the optimal solutions. We first try to solve (3.1) by a search for critical points of the functional J. The task is to find functions $u \in S(\Omega)$ so that the differential $DJ(u)$ of J in the Fréchet sense, given by (3.2), vanishes. Denoting by $S_0(\Omega)$ the set of functions in $S(\Omega)$ vanishing in $\partial\Omega$,

$$S_0(\Omega) = \left\{u : \bar{\Omega} \to \mathbb{R}, \quad u_{|\Omega} \in C^2(\Omega, \mathbb{R}), \text{ s.t. } u_{|\partial\Omega} = 0\right\} \subset S(\Omega),$$

the aim is to find $u \in S(\Omega)$ such that

$$\begin{aligned} DJ(u)v = \int_\Omega \frac{\nabla u \cdot \nabla v}{(1 + |\nabla u|^2)^{\frac{1}{2}}}\, dx = 0 \quad \forall v \in S_0(\Omega), \\ u = \gamma \text{ on } \partial\Omega. \end{aligned}$$

Given a good enough initialization function u_0, we seek a zero of the first derivative of the cost function by means of a Newton algorithm that returns at the k-th iterate a new function u_{k+1} in $S(\Omega)$ such that

$$\begin{aligned} D^2J(u_k)(u_{k+1}, v) - D^2J(u_k)(u_k, v) = DJ(u_k)(v) \quad \forall v \in S_0(\Omega), \\ u_{k+1} = \gamma \text{ on } \partial\Omega. \end{aligned} \tag{3.4}$$

Newton's method strongly depends on the starting point. Therefore, to improve the convergence of the algorithm, the first steps of the algorithm can be devoted to approximating the solution by using a fixed point algorithm before launching the Newton method that will then converge faster once the iterate is close to the solution.

In the first iterations $k \in \{0..K\}$, we thus seek a function $u_{k+1} \in S(\Omega)$ solution of the linear problem

$$\int_{\Omega} \frac{\nabla u_{k+1} \cdot \nabla v}{(1+|\nabla u_k|^2)^{\frac{1}{2}}}\, dx = 0 \quad \forall v \in S_0(\Omega), \qquad u_{k+1} = \gamma \text{ on } \partial\Omega, \tag{3.5}$$

and we switch to the Newton method when the distance between two successive iterations,

$$\left(\int_{\Omega} (u_{k+1} - u_k)^2\, dx\right)^{\frac{1}{2}},$$

is small enough. We stop the Newton method when the above error is below a given tolerance. We write in Code 3.3 the full algorithm combining the fixed point iterations (3.5) and the Newton iterations (3.4).

```
while(err > eps && iter++ < 100)
{
    up[]=u[]; // u_{k-1} ← u_k

    newton = newton | ( err < 0.005); // boolean

    if( ! newton ) // at first iterations
        solve PointFixe(u,v) = DJfp(u,up,v) + on(1,2,3,4,u=gamma)
            ;
    else
        solve Newton(u,v) =
          D2J(up,v,u) - D2J(up,v,up)
        + DJ(up,v)
        + on(1,2,3,4,u=gamma);
    err = sqrt(int2d(Th)((u-up)^2)); // error
    real Ju = J(u);
    cout << iter << " " << newton << " " << err  << " J = "<< Ju
        << endl;
    plot(u,cmm=iter + " err = " + err + " Ju=" + Ju);
}
```

Code 3.3 Plateau problem by fixed point and Newton method (see **minsurf_fixedpoint_Newton.edp**)

Newton's algorithm converges to the solution in few iterations. Nevertheless, the method could appear limited if we would like to add some constraints such as forcing the minimal surface to lie under a given surface. Indeed, an additional constraint would add in the optimality conditions a Lagrange multiplier, which is more complicated to deal with. In such a case, a more user-friendly alternative may be to solve the problem by combining `FreeFEM` with `IpOpt`.

3.3 Solving with FreeFEM Combined with IpOpt

In this section we introduce an additional lower bound function $u_m : \bar{\Omega} \to \mathbb{R}$ and we now consider the optimization problem

$$\min_{u \in S(\Omega)} J(u) \text{ s.t. } \quad u_{|\partial\Omega} = \gamma, \quad u_m \leqslant u. \tag{3.6}$$

In the previous section, we computed the optimal solutions of (3.1) by seeking the critical points of the function J. This is equivalent to solving the first-order optimality conditions which are also sufficient since the problem (3.1) is convex. In this section, the additional lower bound constraint breaks the equivalence between critical points and optimality conditions, which require the introduction of a Lagrange multiplier associated to the constraint. We could apply the first-order optimality conditions and solve them thanks to an augmented Lagrangian method for instance. Here, we rather use the combination of FreeFEM with IpOpt and we discretize the derivatives of the cost function J and of the lower bound constraint. Therefore, in contrast to the previous method, which is typical of the *FOTD* method, since we first wrote the optimality conditions of the problem (3.1) that we have then discretized and solved, we now compute a discretization of the derivatives the functions involved in the problem (3.6) and only after we use an optimization algorithm (here IpOpt). To this aim, we compute a discretized version of $DJ(u)$ and $D^2J(u)$.

```
func real J(real[int] &X)
{
    Vh u; u[]=X;
    return int2d(Th)(sqrt(1.0 + grad(u)'*grad(u))) ;
}

func real[int]  DJ(real[int] &X)
{
    Vh u; u[]=X;
    varf vg(uu,v) = int2d(Th)((grad(u)'*grad(v))/sqrt(1.0 + grad(
        u)'*grad(u))) ;
    real[int] G= vg(0,Vh);
    return G;
}

matrix H;  //global variable for matrix, otherwise => seg fault
    in Ipopt
func matrix  D2J(real[int] &X)
{
    Vh u; u[]=X;
    varf vH(v,w) = int2d(Th)( (grad(w)'*grad(v))/sqrt(1.0 +grad(u
        )'*grad(u))
      - (grad(w)'*grad(u))*(grad(v)'*grad(u))*(1.0 + grad(u)'*grad
         (u))^-1.5 ) ;
    H = vH(Vh,Vh);
```

```
    return H;
}
```

Code 3.4 Optimal surface (see **minsurf_Ipopt.edp**)

We directly define the additional constraints in lower and upper bounds of the optimization variable X. For the problem (3.1), we force in Code 3.5 the function u to be equal to γ on the boundary $\partial\Omega$ by introducing two arrays l_b and u_b so that

$$l_b \leqslant X \leqslant u_b$$

with $l_b[i]$ and $u_b[i]$ very large if the index i matches with a degree of freedom of an interior finite element and $l_b[i]$ and $u_b[i]$ equal to γ if the index i matches with the degree of freedom of a boundary finite element.

```
varf OnGamma(u,v)  = on(1,2,3,4,u=1);
func g = cos(pi*x)*cos(pi*y);
Vh OnG; Vh gh=g;
OnG[]=OnGamma(0,Vh,tgv=1); // 1 on Gamma
Vh lb = OnG!=0 ? gh : -1e19 ; //
Vh ub = OnG!=0 ? gh :  1e19 ; //
```

Code 3.5 Boundary constraint

We define the additional constraint stated in the problem (3.6) by adding the following two lines in the code in order to modify the lower bound variables accordingly for the function

$$u_m(x, y) = 3 - \left(10\left((x-0.5)^2 + (y-0.5)^2\right)\right)^2.$$

```
Vh clb = 3-square(10*(square(x-x0)+square(y-y0))); // lower
    constraint
if(constraint) lb = max(lb,clb);
```

It only remains to run `IpOpt`:

```
IPOPT(J,DJ,HJ,u[],lb=lb[],ub=ub[],tol=1.e-15);
```

In this example, we first computed a discretized version of the derivatives of the cost function, what can be referred to as a hybrid method in Diagram 1.2. A full *FDTO* method would compute the derivatives of a discretized version of the cost function.

3.4 Automatic Differentiation Alternative

In the two previous sections, we computed the continuous derivatives of the involved functions and then discretized them according to our finite element space's choice. From the numerical point of view, having in mind a gradient descent method as an optimization algorithm, it may sometimes be better to compute the true numerical gradient, i.e., compute the derivatives of a discretized version of the cost and constraint functions. This is why we focus in this section on an automatic differentiation alternative based on the software AMPL. We are going to see that we have to choose adequately the discretization space, otherwise convergence of the algorithm may fail. In a first attempt, as in the previous sections, we choose $\mathbb{P}_1$ finite elements for the function u. The cost function brings the spatial derivatives of u which have to be discretized too. When we write in FreeFEM the cost function

```
func real J(real[int] &X)
{
    Vh u; u[]=X;
    return int2d(Th)(sqrt(1.0 + grad(u)'*grad(u))) ;
}
```

Here, $dx(u)$ and $dy(u)$ discretizations are hidden in the quadrature formula int2d. Automatic differentiation is not directly available in FreeFEM. Therefore, we have to export the discretization in AMPL and we thus have to exhibit this formula in order to well observe the same solution. To this aim, we are going to develop the cost function computation in the Code 3.6 and comment the computing steps.

```
param mV integer >=0; # ndof for u discretization
param mP integer >=0; # ndof for dx(u) and dy(u) discretizations

set indexV = 0..mV-1;
set indexP = 0..mP-1;

set indexPx dimen 2;
set indexPy dimen 2;
set indexM dimen 1;
set indexG dimen 1;
set indexUE dimen 1;

param Px{indexPx}; # matrix of operator dx
param Py{indexPy}; # matrix of operator dy
param M{indexM}; # mass matrix
param ue{indexUE};

var u{indexV};

minimize cost : sum{i in indexM}( M[i]*sqrt(1.0+ (sum{(i,j) in
    indexPx}( Px[i,j]*u[j] ))**2 + (sum{(i,j) in indexPy}(
    Py[i,j]*u[j] ))**2));
```

```
subject to gamma_inf{i in indexG} : u[i]<=ue[i] + 1.e-19; #
subject to gamma_sup{i in indexG} : u[i]>=ue[i] - 1.e-19; #

option solver ipopt;
option ipopt_options"max_iter=1000 tol=1.e-15
    linear_solver=mumps";
```

Code 3.6 File "surfmin.mod" (see **minsurf.{dat,mod,run}**)

We compute the matrices involved in Code 3.6 in FreeFEM by introducing two finite element spaces: V_h is used to discretize the optimization variable u, while P_h is used to discretize $dx(u)$ and $dy(u)$. Indeed, given u approximated with $\mathbb{P}_1$ finite elements, its spatial derivatives are thus discretized with $\mathbb{P}_0$ finite elements without approximation error. Therefore the two matrices P_x and P_y map the finite element space V_h to the finite element space P_h.

```
fespace Vh(Th,P1);
fespace Ph(Th,P0);

matrix Px = interpolate(Ph,Vh,t=0,op=1);
matrix Py = interpolate(Ph,Vh,t=0,op=2);

varf mass(u,v) = int2d(Th)(v);
real[int] M = mass(0,Ph);
```

Code 3.7 Matrices needed in "surfmin.mod"

3.5 Example and Remarks

We illustrate the previous methods by choosing

$$\gamma(x, y) = \cos(\pi x)\cos(\pi y),$$

defined on the square $\Omega = [0, 1]^2$. In a first attempt, we discretize the unknown u with $\mathbb{P}_1$ finite elements and we plot in Fig. 3.1 the solution returned by the three methods.

Indeed, solving the first-order optimality conditions either with the automatic differentiation method or by using IpOpt in FreeFEM, the solution found is almost identical. To convince ourselves of this, we plot in Fig. 3.2 the L^2 error between the solutions and we observe that they are similar to within 10^{-7}.

The fact that the solutions are identical comes from the choice of the discretization space, which allows us to obtain a discretized cost function without interpolation error. Indeed, we have taken $\mathbb{P}_1$ elements for u so that its spatial

Fig. 3.1 Solution returned by the three methods

Fig. 3.2 Error between different solutions for $\mathbb{P}_1$ discretization

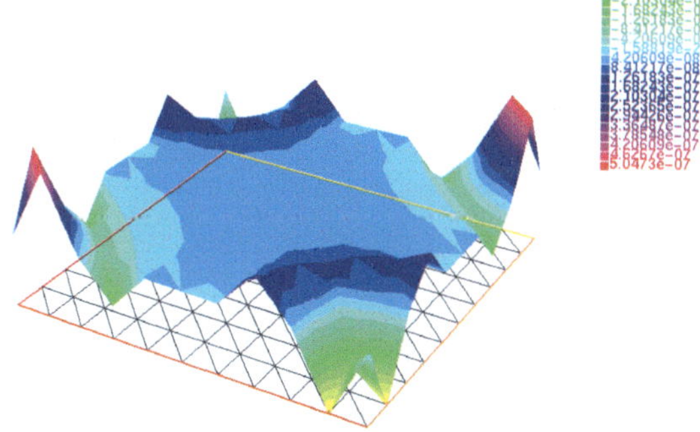

derivatives can be approximated by $\mathbb{P}_0$ elements, which are constant functions on each triangle of the mesh. Hence, the quadrature formula used to compute the cost function involves constant values of the function $\sqrt{1+|\nabla u|^2}$ on each triangle. However, if the function approximated inside the root is no longer constant, we will make approximation errors and the three methods will not return exactly the same solution. For example, let us now take as discretization space the $\mathbb{P}_2$ elements for u and let us plot the solution in Fig. 3.3. Here, the spatial derivatives of u are discretized according to discontinuous $\mathbb{P}_1$ elements. If we try to discretize them with continuous $\mathbb{P}_1$ elements, then we make an approximation error which is prohibitive for the convergence of the algorithm (Fig. 3.4).

These two numerical examples illustrate that the choice of the discretization's space matters to ensure the convergence of the algorithm and for direct and indirect methods to lead to the same result.

Fig. 3.3 Solution of Plateau's problem with $\mathbb{P}_2$ elements: (a) with AD (AMPL); (b) with IpOpt and FreeFEM

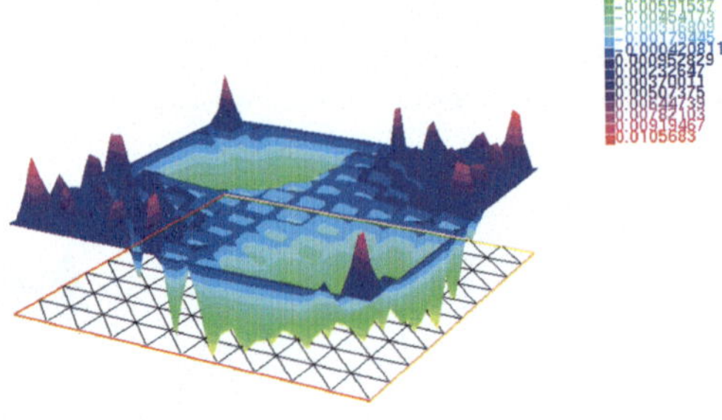

Fig. 3.4 Error between different solutions for $\mathbb{P}_2$ discretization

In a last example, we illustrate how IpOpt combined with FreeFEM is standing out. We now force the solution to be under a given function as an additional constraint. We consider the problem

$$\min_{u\in S(\Omega)} J(u) \text{ such that } u_{|\partial\Omega} = \gamma, \quad u \leqslant u_M \tag{3.7}$$

with

$$\gamma(x, y) = \cos(\pi x)\cos(\pi y),$$

and

$$u_M(x, y) = 3 - (10((x - 0.5)^2 + (y - 0.5)^2))^2.$$

Solving optimality conditions would require to add and manage a Lagrange multiplier for the obstacle constraint. Besides, automatic differentiation proved to be a relevant solution if we discretize with $\mathbb{P}_1$ elements, but it is still limited for other discretization choices. Nevertheless, IpOpt combined with FreeFEM manages a

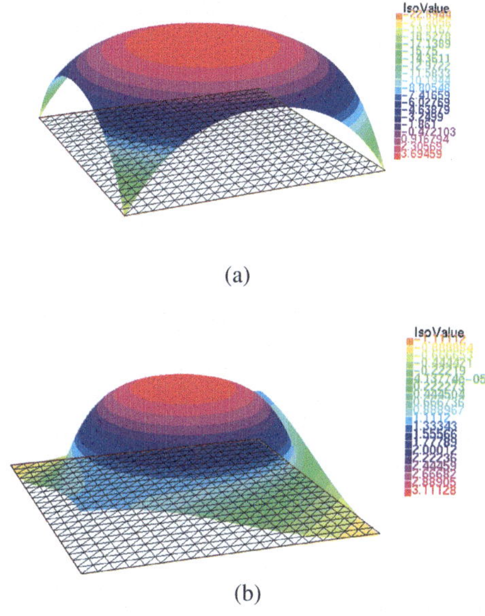

Fig. 3.5 Solution of Plateau's problem with $\mathbb{P}_2$ elements: (**a**) obstacle u_M; (**b**) solution with obstacle

large choice of discretization spaces and enables to take into account additional constraint as inputs. This is why this solution turns out to be an effective and interesting one. We plot in Fig. 3.5 the solution of the Plateau problem with an obstacle constraint u_M.

All numerical codes developed in this chapter are downloadable at `FreeFEM`'s website.

References

1. Jesse Douglas. 1931. Solution of the problem of plateau. *Transactions of the American Mathematical Society* 33(1): 263–321.
2. Tibor Radó. 1993. *On the problem of plateau*, 1st edn. Number VII, 109 in Ergebnisse der Mathematik und Ihrer Grenzgebiete. Springer:Berlin/Heidelberg.

Chapter 4
Shape Optimization with `FreeFEM`

Abstract In this chapter, we address a less classical problem that illustrates the user-friendly yet effective framework provided by `FreeFEM` to solve a shape optimization problem with mesh modifications combined with the powerful interior point method `IpOpt`. Based on the mathematical framework introduced in Chap. 1 and following the numerical tools illustrated in Chap. 2, we consider the optimal design problem of a microfluidic swimmer's membrane consisting of maximizing its velocity in a given direction. We treat the membrane as a moving lower boundary of a domain where a Stokes fluid evolves and we want to maximize the fluid velocity at the free surface, by imposing constraints on the length and curvature of the boundary.

The examples presented in the previous sections were rather classical and aimed at illustrating several numerical methods and providing some numerical templates.

In this section, our objective is to highlight other features of `FreeFEM` on a more challenging example, namely, the problem of designing the best possible shape for micro-swimmers in a fluid, in order to lead them in a given direction. At the interface between fluid-structure interaction and control theory, this problem has been studied in various forms in the literature (see [2, 10]). Here, as illustrated in Fig. 4.1, we assume that the micro-swimmers have a membrane Γ that oscillates periodically at some fixed frequency to generate fluid motion. To account for the periodicity of the membrane motion, the fluid domain is assumed to be a flat torus obtained by bending the square along the y direction so that Σ_2 merges with Σ_4 (torus); the lower boundary Γ is assumed to move at speed v_D so that viscous forces in Γ imply fluid motion (see Fig. 4.1b). For the ease of study, we assume that Γ is the graph of a C^2 function (no overlap), moreover satisfying some physical constraints on the second derivative.

In a more complicated model, overlap could be allowed by modifying the boundary with two-dimensional deformation vector fields.

We perform our study in the reference frame of the moving boundary Γ (velocity v_D), which induces the vertical motion of the fluid observed in Fig. 4.3.

F. Hecht et al., *PDE-constrained optimization within FreeFEM*, SpringerBriefs on PDEs and Data Science, https://doi.org/10.1007/978-981-95-7049-2_4

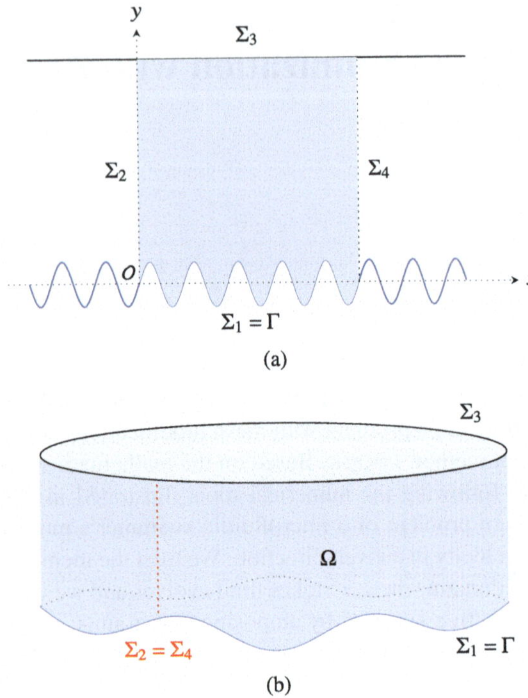

Fig. 4.1 Fluid domain: (**a**) initial square; (**b**) torus after periodization

4.1 Boundary and Domain Parametrization

Let Ω_0 denote the initial torus. Given a function

$$f : [0, 1] \to [0, 1]$$

such that $f(0) = f(1)$, we parametrize the bottom boundary as

$$\Gamma = \{(x, f(x)) \mid x \in [0, 1]\},$$

so that Ω_0 is modified to obtain the fluid domain Ω under the action of the vector field

$$\theta(x, y) = (x, y + (1 - y)f(x)).$$

As for the numerical approach, although the flow problem introduced below is expressed on the modified cylinder Ω, we discretize the domain with a square mesh such that periodicity will be ensured by considering periodic finite elements. We then introduce a triangulation of Ω (see Fig. 4.2b) by modifying the previous square mesh (see Fig. 4.2a) under the action of the vector field θ thanks to the command `movemesh`. As the control f acts on the 1D boundary, we introduce a

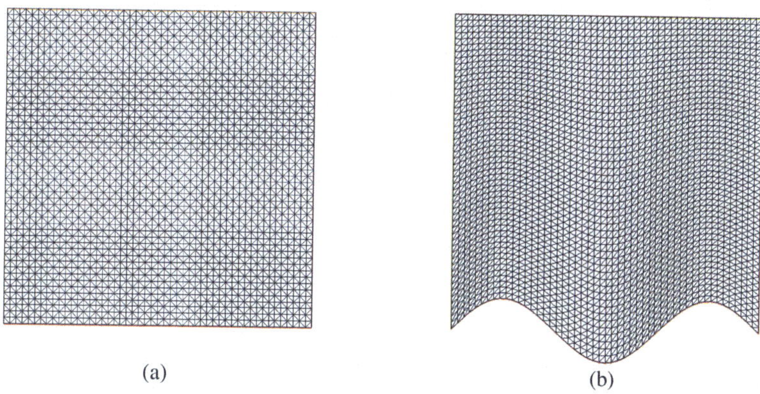

Fig. 4.2 Fluid domain mesh: (**a**) initial; (**b**) modified for $f(x) = 0.1\sin(3\pi x)$

one-dimensional mesh of the boundary which will be necessary for the computation of the gradient. Each iteration of the optimization process will involve a modification of both the square mesh and the one-dimensional mesh along Γ. Numerically, in the Code 4.1, we introduce several meshes in addition to the initial square mesh of Ω_0 thanks to the command `extract`:

```
int[int] ll=[1];
mesh Th0 = square(NX,NX*H,[x,y*H]); // initial square mesh
func fclab = (x>0.999)*2 + (x<0.001)*1; // Bordcrs label
ThL = extract(Th0,refedge=ll);
ThL = change(ThL, flabel = fclab); // (0,1) straight mesh with
    label 1, 2

mesh Th = Th0; // mesh to be modified at each function call to
    get Ω
meshL ThC = ThL; // mesh to be modified at each function call to
    get Γ
```

Code 4.1 1D finite element space and movemesh command (see **shape_microswimmer.edp**)

Figure 4.2 shows the initial square mesh of Ω_0 and its modification to obtain the mesh of Ω for a particular function.

The reference frame of the moving boundary follows a rectilinear uniform motion with speed v_D in the x-direction. Let $u = (u_1, u_2)$ be the velocity of the fluid and let p be the pressure. We consider the Stokes flow equations on the modified domain Ω:

$$-2\mu\nabla\cdot\varepsilon(u) + \nabla p = 0 \quad \text{in } \Omega \tag{4.1}$$

$$\operatorname{div}(u) = 0 \quad \text{in } \Omega \tag{4.2}$$

$$\sigma(u, p)\mathbf{n} = 0 \quad \text{on } \Sigma_3 \tag{4.3}$$

$$u = \begin{pmatrix} 0 \\ f' \end{pmatrix} \quad \text{on } \Gamma \tag{4.4}$$

where $\mathbf{n}$ is the outer normal vector, μ is the kinematic viscosity,

$$\varepsilon(u) = \frac{1}{2}\left(\nabla u + \nabla u^T\right)$$

is the symmetric gradient of u and

$$\sigma(u, p) = 2\mu\varepsilon(u) - p\,\mathrm{Id}$$

is the Cauchy stress tensor. Assuming that

$$f \in H^2(0, 1) \cap H_0^1(0, 1),$$

Ω has at least a $C^{1,1}$ boundary by Sobolev injections. Existence of solutions

$$(u, p) \in H^1(\Omega) \times L^2(\Omega)$$

of the Stokes system (4.1, 4.2, 4.3, 4.4) follows from [4, Theorem IV.5.2]. More details on existence and regularity of solutions in less regular domains can be found in [3, 8].

Remark 4.1 Equation (4.1) stands for the balance law between viscous and pressure forces, while (4.2) stands for the incompressibility of the fluid. We usually add an integral constraint $\int_\Omega p(x)\,dx = 0$ on the pressure in order to guarantee the uniqueness of the solution (u, p), which is not needed here since the condition (4.3) yields the pressure on Σ_3. Equations (4.3) and (4.4) respectively imply that the fluid is free of surface forces at the top boundary Σ_3 and a no-slip boundary condition on Γ, namely, the fluid and the solid have the same speed at the interface. For micro-swimmers, inertia effects play no role and the motion is entirely determined by the friction forces encompassed in (4.4) and therefore by the shape of Γ.

Given a boundary shape

$$f(x) = 0.1\sin(3\pi x)$$

moving to the right ($v_D = -1$), we plot the corresponding velocity in Fig. 4.3. Near the minimum of f, we observe that the slope of the Γ boundary generates a semblance of a vortex and a fluid velocity that increases with the variation of the slope. The mixed boundary conditions (4.3) and (4.4) are treated in the variational formulation by introducing the *state* finite element space

$$V = \{(u, p) \in H^1(\Omega)^2 \times L^2(\Omega) \mid u = \begin{pmatrix} 0 \\ f' \end{pmatrix} \text{ on } \Gamma\},$$

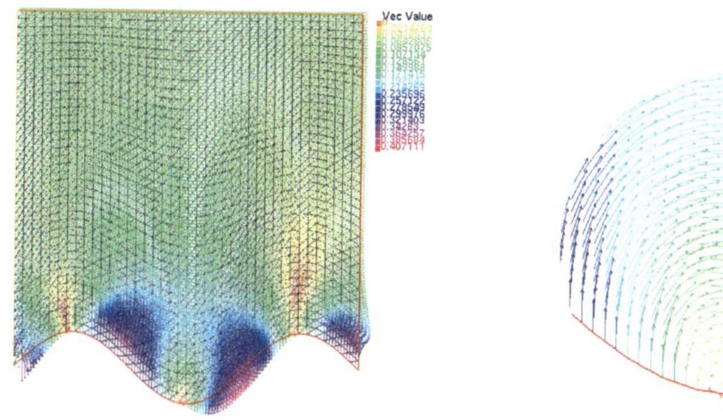

Fig. 4.3 Velocity u for $f(x) = 0.1\sin(3\pi x)$ and $v_D = -1$

so that $(u, p) \in V$ is a weak solution of the Stokes system if and only if

$$\int_\Omega (2\mu\varepsilon(u) : \varepsilon(v) - q\,\mathrm{div}(u) - p\,\mathrm{div}(v))\,dx = 0 \qquad \forall(v, q) \in V_0 \tag{4.5}$$

where

$$V_0 = \{(v, q) \in H^1(\Omega)^2 \times L^2(\Omega) \mid v = 0 \text{ on } \Gamma\}.$$

As in section "PDE-constrained Optimization", the solution (u, p) belongs to an affine space. The finite element space is chosen such that the discretized linear system corresponding to (4.1, 4.2) is invertible. For example, one can choose the Lagrangian elements $\mathbb{P}_2$ for the velocity u and the Lagrangian elements $\mathbb{P}_1$ for the pressure p, which are known to imply an LBB condition (see [6, Chap. 2]) that guarantees this invertibility property. The variational formulation (4.5), whose state and test functions are defined on the modified domain Ω, is solved numerically in a finite element space with periodic boundary conditions via the command

```
fespace Wh(Th,[P2,P2,P1],periodic=[[2,y],[4,y]]); //
    [u1,u2,p], Σ2 = Σ4.
Wh [u1,u2,p], [v1,v2,q];
```

with the following macros related to the variational formulations

```
macro SGrad(u,v) [[dx(u),0.5*(dx(v)+dy(u))],[0.5*(dx(v)+dy(u)),dy
    (v)]]
                     // ε(u)
macro div(u1,u2) (dx(u1)+dy(u2)) // div(u)
```

to solve (4.1, 4.2, 4.3, 4.4) with the state equation macro stated in Code 4.2.

```
macro stokes() { // State equation's macro (4.1, 4.2, 4.3, 4.4)
  solve Stokes( [u1,u2,p], [v1,v2,q] ) =
      int2d(Th)(2*mu*(SGrad(u1,u2):SGrad(v1,v2)) - div(u1,u2)*q)
    - int2d(Th)(div(v1,v2)*p)
    + on(1,u1=0,u2=vdent*gm)} //
```

Code 4.2 Stokes state equation

The 1D control $f \in H^2(0,1) \cap H^1(0,1)$ is discretized with one-dimensional finite elements $\mathbb{P}_1$ into fm. Therefore, its exact numerical derivative **dx**(fm) is discretized with finite elements $\mathbb{P}_0$, i.e. it is piecewise constant. However, the boundary condition **on**(1,u1=0,u2=vd*gm) of the Code 4.2 implies that u2 which is continuous, must be equal to gm which must therefore be continuous. In an effort to minimize approximation errors, we approximate in the Code 4.3 the initial $\mathbb{P}_0$ discretization of f' using the projection on $L^2(0,1)$ with $\mathbb{P}_1$ finite elements.

```
func real[int] L2regul(real[int] &X)
{
    WhL f,g,cdf;
    f[] = X;
    solve l2regul(cdf,g) = int1d(ThL)(cdf*g) // L²(Ω) projection
        of dx(f) (P0)
                         - int1d(ThL)(dx(f)*g); // with P1
                           elements
    return cdf[];
}
```

Code 4.3 Regularization of f' with L^2 projection with $\mathbb{P}_1$ finite elements

4.2 Shape Optimization Problem

In Sect. 4.1, the problem consists in maximizing the velocity of the fluid in the direction $\iota = \binom{1}{0}$ with respect to the boundary Γ, by considering the cost functional

$$\widehat{J}(f) = -\int_{\Sigma_3} u \cdot \iota \, ds \tag{4.6}$$

where the pair (u, p) satisfies (4.1, 4.2, 4.3, 4.4). Some additional constraints on f are added to ensure both existence of optimal solutions and a better convergence of the optimization algorithm. The set $\mathcal{U}_{ad}$ entails *physical* constraints:

- The volume constraint

$$\int_0^1 f(x)\, dx = 0$$

implies that the domain Ω keeps a constant volume in time. It is formulated numerically by introducing the vector `Cvol` that represents the linear form

$$g \mapsto \int_0^1 g(x)\, dx$$

in the $\mathbb{P}_1$ finite element basis:

```
varf varCvol(f,g) = int1d(ThL)(1*g); // ∫_0^1 g(x) dx
real[int] Cvol = varCvol(0,WhL);
```

Code 4.4 Volume constraint

- Bound constraints on the first derivative of f

$$|f'(x)| \leqslant M_1 \quad \forall x \in (0, 1)$$

  ensure existence of optimal solutions. This constraint is numerically carried out by using the interpolation matrix of the derivative operator from $\mathbb{P}_1$ finite elements to $\mathbb{P}_0$ finite elements defined in Code 4.6.
- To ensure existence of a solution $(u, p) \in H^1(\Omega) \cap L^2(\Omega)$ of the Stokes system, we assume that $f \in H^2(\Omega)$ so that Ω has a $C^{1,1}$ boundary and $f' \in H^1(0, 1)$, and we assume in addition that there is a bound constraint on the second derivative of f:

$$|f''(x)| \leqslant M_2 \quad \forall x \in (0, 1).$$

  This curvature constraint, which is physical, is important to ensure well-posedness and we observe that it induces a better convergence of the optimization algorithm.

The optimization problem (4.1, 4.2, 4.3, 4.4, 4.6) is formulated as

$$\min_{f \in \mathcal{U}_{ad}} \widehat{J}(f)$$

where $\mathcal{U}_{ad}$ is the subset of $U = H^2(0, 1)$ defined by

$$\mathcal{U}_{ad} = \Big\{ f \in H^2(0, 1) \cap H_0^1(0, 1) \mid \int_0^1 f(x)\, dx = 0, \\ |f'(x)| \leqslant M_1,\ |f''(x)| \leqslant M_2 \ \text{ for a.e. } x \in (0, 1) \Big\}.$$

4.3 Sensitivity Analysis

The function f acts on both the shape of the Ω domain (f' is involved in the boundary condition (4.4)) and the solution (u, p) of the Stokes system (4.1, 4.2, 4.3, 4.4). If the domain Ω were fixed, we could compute the derivative of the reduced cost function directly using the adjoint representation introduced in section "PDE-constrained Optimization", but here the shape deformations of Γ must also be taken into account. To this end, we come back to the variational calculus approach and we write a sensitivity analysis to express the derivative.

Let $f \in \mathcal{U}_{ad}$. Let us compute the shape derivative of the reduced cost function in the direction $g \in U$ by taking t small enough so that $f + tg \in \mathcal{U}_{ad}$. We define

$$\Omega_t = (\mathrm{id} + t\phi)(\Omega)$$

where

$$\phi(x, y) = \begin{pmatrix} 0 \\ \frac{1-y}{1-f(x)} g(x) \end{pmatrix}$$

is the vector field deforming Ω to Ω_t with respect to the small perturbation g. We define the real-valued function

$$F : t \mapsto F(t) = \widehat{J}(f + tg)$$

and we compute its Fréchet derivative $F'(0)$ which gives the reduced cost function derivative

$$F'(0) = \langle D\widehat{J}(f), g \rangle_{U',U}.$$

According to usual methods for the computation of derivatives of solution of a PDE with respect to the domain (see [1] and [7, Chap. 5]) we introduce the material derivative $(\tilde{u}, \tilde{p})$ solution of the linearized system

$$-2\mu\nabla \cdot \varepsilon(\tilde{u}) + \nabla\tilde{p} = 0 \quad \text{in } \Omega \tag{4.7}$$

$$\mathrm{div}(\tilde{u}) = 0 \quad \text{in } \Omega \tag{4.8}$$

$$\sigma(\tilde{u}, \tilde{p})\mathbf{n} = 0 \quad \text{on } \Sigma_3 \tag{4.9}$$

$$\tilde{u} = v_D \begin{pmatrix} 0 \\ g' \end{pmatrix} - \frac{\partial}{\partial n}\left(u - v_D \begin{pmatrix} 0 \\ f' \end{pmatrix}\right) \times \phi \cdot \mathbf{n} \quad \text{on } \Gamma. \tag{4.10}$$

The derivative of the reduced cost function $\widehat{J}$ is

$$\langle D\widehat{J}(f), g\rangle_{U',U} = \int_{\Sigma_3} \tilde{u} \cdot \iota \, ds.$$

Two terms appear in (4.10), respectively coming from the boundary condition $v_D \begin{pmatrix} 0 \\ f' \end{pmatrix}$ and from the domain variation vector field ϕ in (4.10). The first term stands for the variation of u with respect to the boundary condition (4.4), assuming Ω fixed, while the second one stands for the mesh deformation variation. By (4.10), we thus get $\tilde{u}$ along Γ, under the influence of both terms, while the knowledge of $\tilde{u}$ along Σ_3 is required to compute $\langle D\widehat{J}(f), g\rangle_{U',U}$. The adjoint vector $(v, q) \in H^1(\Omega)^2 \times L^2(\Omega)$ is defined as the solution of

$$-2\mu\nabla \cdot \varepsilon(v) + \nabla q = 0 \quad \text{in } \Omega \tag{4.11}$$

$$\text{div}(v) = 0 \quad \text{in } \Omega \tag{4.12}$$

$$\sigma(v, q)\mathbf{n} = \iota \quad \text{on } \Sigma_3 \tag{4.13}$$

$$v = 0 \quad \text{on } \Gamma. \tag{4.14}$$

Remark 4.2 The adjoint system (4.11, 4.12, 4.13, 4.14) is similar to the one we would get from the Stokes problem (4.1, 4.2, 4.3, 4.4) considered on a fixed domain Ω with a Dirichlet boundary control.

Following the variational formulation (4.5) for the numerical solving of the state equation, we solve in Code 4.5 the adjoint system (4.11, 4.12, 4.13, 4.14).

```
macro adjoint() {
    solve StokesAdjoint( [v1,v2,q],[w1,w2,g] ) =
          int2d(Th)( 2*mu*(SGrad(v1,v2):SGrad(w1,w2)) - div(w1,w2
              )*q )
        - int2d(Th)( div(v1,v2)*g )
        - int1d(Th,3)(w1) // Neumann condition (4.13)
        + on(1,v1=0,v2=0); // Dirichlet condition (4.14).
} //
```

Code 4.5 Stokes adjoint system

Using the boundary condition (4.13), we have

$$\langle D\widehat{J}(f), g\rangle_{U',U} = \int_{\Sigma_3} \tilde{u} \cdot \iota \, ds = \int_{\Sigma_3} \sigma(v, q)\mathbf{n} \cdot \tilde{u} \, ds.$$

The function $\tilde{u}$ is known along Γ thanks to (4.10), involving the effects of f on the derivative. To express the derivative with respect to f, we use the adjoint system (4.11, 4.12, 4.13, 4.14), in relationship with the expression of $\sigma(v,q)\mathbf{n}\cdot\tilde{u}$ on Γ, and on Σ_3 by making an integration by parts:

$$\begin{aligned}\int_\Omega (-2\mu\nabla\cdot\varepsilon(u)+\nabla p)\cdot w\,dx &= \int_\Omega (2\mu\varepsilon(u):\varepsilon(w)-p\mathrm{div}(w))\,dx\\ &\quad -\int_{\Gamma\cup\Sigma_3}\sigma(u,p)\mathbf{n}\cdot v\,ds.\end{aligned}\tag{4.15}$$

Applying (4.15) to the solution (u,p) of Stokes system, to the solution $(\tilde{u},\tilde{p})$ of the linearized system and to the solution (v,q) of the adjoint system, we finally compute the reduced cost function derivative as

$$\langle D\widehat{J}(f),g\rangle_{U',U} = \int_\Gamma \sigma(v,q)\mathbf{n}\cdot\left(v_D\begin{pmatrix}0\\g'\end{pmatrix}-\frac{\partial}{\partial n}\left(u-v_D\begin{pmatrix}0\\f'\end{pmatrix}\right)\phi\cdot\mathbf{n}\right)ds \tag{4.16}$$

where

$$\mathbf{n}=\frac{1}{(1+f'^2)^{\frac{1}{2}}}\begin{pmatrix}f'\\-1\end{pmatrix}.$$

We identify the linear form (4.16) with a gradient expressed in $U = H^2(0,1)$ by finding, for a $H^2(0,1)$-inner product to be defined in accordance with the duality pairing $\langle\cdot,\cdot\rangle_{U',U}$, the solution $\rho\in H^2(0,1)$ of

$$(\rho,g)_{H^2(0,1)} = \int_\Gamma (\phi_1(x,y)g'(x)+\phi_2(x,y)g(x))\,ds \qquad \forall g\in H^2(0,1), \tag{4.17}$$

with

$$\phi_1 = v_D\sigma(v,q)\mathbf{n}\cdot\begin{pmatrix}0\\1\end{pmatrix},$$

$$\phi_2=\sigma(v,q)\mathbf{n}\cdot\frac{\partial u}{\partial n}\times\phi\cdot\mathbf{n}-v_D\sigma(v,q)\mathbf{n}\cdot\begin{pmatrix}0\\1\end{pmatrix}\times f''\mathbf{n}_1\phi\cdot\mathbf{n},$$

where $\mathbf{n}_1$ is the x-axis component of $\mathbf{n}$. The above quantities are stored numerically in data types **func** The state (u, p) and the adjoint (v, q) are discretized with finite elements $\mathbb{P}_2 \times \mathbb{P}_1$ and the control f is discretized with 1D finite elements $\mathbb{P}_1$. Since $f \in H^2(0, 1)$, we do not consider here conformal transformations for U but rather nonconformal transformations, in agreement with the general mathematical framework given in the remark 1.7. Moreover, ϕ_1 and ϕ_2 are defined on the modified mesh in (4.17) and their finite element approximation in the Code 4.7 is discontinuous since they involve derivatives of f. Therefore, we introduce below several 1D finite element spaces defined respectively on the square and modified meshes `ThL` and `ThC` to interpolate ϕ_1 and ϕ_2 with finite elements $\mathbb{P}_1$ on $(0, 1)$ (see Code 4.7).

```
fespace WhL(ThL,P1); // F.E. functions on straight mesh: f
fespace PhL(ThL,P0); // F.E. functions on straight mesh: f'
fespace WhC(ThC,P1); // F.E. functions on curve mesh
```

Let us give some details: `WhL` is a finite element space used to discretize f and to write the matrix of the scalar product on $H^2(0, 1)$ which is required for gradient interpolation; `PhL` is a finite element space with which we compute the first and second derivatives of f that appear in the constraints and are required to compute the resulting matrix of the second-order terms involved in the $H^2(0, 1)$-inner product. The matrix corresponding to the first derivative operator is constructed using the operator **dx** which thus allows to express the derivative of a finite element function $\mathbb{P}_1$ with finite elements $\mathbb{P}_0$. For the second derivative, we construct by hand the matrix of jumps of the first order derivatives of the finite element basis $\mathbb{P}_1$ `WhL` (see Code A.1 further).

```
matrix mDx  = interpolate(PhL,WhL,t=0,op=1); // ∂x operator: P1
    to P0
matrix mDxx;
MatJumpofDx(WhL,ThL,mDxx); // returns the jumps of mDx (see
    Code A.1)
```

Code 4.6 Matrices of first and second derivatives of f

Finally, `WhC` allows us to numerically compute the integral along the curve boundary Γ involved in (4.17). In Code 4.7, we compute a regularization of ϕ_1 and ϕ_2 by performing a projection onto $L^2(\Omega)$ with $\mathbb{P}_1$ finite elements on $(0, 1)$.

```
func real[int] L2regulphi(int indexphi)
{
    WhC vC,phiC;
    WhL phiL;
    func phi = (1.0-y)/(1.0-fm);
    func nx = dx(fm)*(1.0+dx(fm)^2)^(-0.5);
    func ny = -1.0*(1.0+dx(fm)^2)^(-0.5);
    if (indexphi == 1){ // for ϕ1 non continuous
        func phi1 = vdent*sigman(v1,v2,q,nx,ny)'*[0.0,1.0];
        solve l2regulphi(phiC,vC) = int1d(ThC)(phiC*vC) - int1d(
            ThC)(phi1*vC);
        }
    if (indexphi == 2){ // for ϕ2 non continuous
        func phi2 = -1.0*phi*ny*sigman(v1,v2,q,nx,ny)'*[dx(u1)*nx
            +dy(u1)*ny,dx(u2)*nx+dy(u2)*ny]
                +vdent*c2gm*nx*phi*ny*sigman(v1,v2,q,nx,ny)
                    '*[0,1];
        solve l2regulphi(phiC,vC) = int1d(ThC)(phiC*vC) - int1d(
            ThC)(phi2*vC);
        }
    phiL = phiC;
    return phiL[];
}
```

Code 4.7 L^2 regularization for ϕ_1 and ϕ_2

The **macro** `sigman(v1,v2,q,nx,ny)` stands for the vector $\sigma(v, q)\mathbf{n}$. This interpolation is necessary because we cannot accurately interpolate the discontinuous approximations of ϕ_1 and ϕ_2 defined on Γ (given by `phi1` and `phi2` in Code 4.7) with continuous $\mathbb{P}_1$ finite elements on $(0, 1)$. Once this regularization is done, the interpolation from Γ to $(0, 1)$ is straightforward in `FreeFEM` by typing `phi1L = phi1C`. We next compute the gradient based on the $H^2(0, 1)$-inner product chosen with the macro in Code 4.8.

```
macro gradInterp(){
    varf linearform(u,v) = int1d(ThL)(-dx(phi1L)*v + phi2L*v); //
        (4.16)
    real bdJ = linearform(0,WhL);
    real[int] dJ = MH2^-1*bdJ; // Gradient in H^2(0,1)
} //
```

Code 4.8 Stokes gradient's interpolation

Remark 4.3 Numerically, we observe that the algorithm is converging in a much better way when we compute the gradient by interpolating the linear form (Code 4.8)

$$(\rho, g)_{H^2(0,1)} = \int_{0^1} (-\phi_1'^L(x) + \phi_{2L}(x))g(x)\,dx \qquad \forall g \in H^2(0,1),$$

$$\text{versus } (\rho, g)_{H^2(0,1)} = \int_{0^1} (\phi_{1L}(x)g'(x) + \phi_{2L}(x)g(x))\,dx \qquad \forall g \in H^2(0,1),$$

where ϕ_{1L} and ϕ_{2L} are respectively the interpolations of ϕ_1 and ϕ_2 with $\mathbb{P}_1$ finite elements on $(0, 1)$, computed in Code 4.7.

The matrix representing the inner product of $H^2(0,1)$ is constructed below with the matrix representing the inner product of $H^1(0,1)$ and the one giving the jumps of the first order derivatives of the finite element basis (see the Code 4.6).

```
// H^2(0,1)-inner product
varf scalarH1(u,v) = int1d(ThL)(u*v+dx(u)*dx(v));
matrix MH1 = scalarH1(WhL,WhL); // (ρ,g)_{H^1(0,1)}
matrix mDxxL = mDxx'*mDxx;
matrix MH2 = MH1 + mDxxL; // (ρ,g)_{H^2(0,1)}
set(MH2,solver=sparsolver);
```

Code 4.9 $H^2(0,1)$-inner product's construction

Remark 4.4 As stated in [5, 9], we may consider a weighted version of the usual $H^2(0,1)$-inner product

$$(f, g)_{H^2(0,1)} = \int_0^1 (\alpha^2 f''g'' + f'g' + fg)\,dx,$$

with $\alpha > 0$ to be tuned depending on the mesh.

4.4 Codes and Results

We finally write the complete algorithm with the several macros and functions defined above and we first initialize the numerical framework based on Sect. 4.1 to solve (4.1, 4.2, 4.3, 4.4).

```
load "msh3"
load "gsl"
load "ff-Ipopt"

bool hotrestart = 1;
verbosity=0;
real HUB = 10.0; // |f''(x)| ⩽ M2
real CUB = 0.5; // |f'(x)| ⩽ M1
real XUB = 0.2;
real H = 1; //
int NX = 25; // boundary mesh size
real vdent = -1.0; // boundary's speed

int[int] ll=[1];
mesh Th0 = square(NX,NX*H,[x,y*H]);
func fclab = (x>0.999)*2 + (x<0.001)*1; // Borders label
meshL ThL = extract(Th0,refedge=ll);
ThL = change(ThL, flabel = fclab); // (0,1) Straight Mesh with
    label 1, 2

mesh Th = Th0;
meshL ThC = ThL; // Curve Mesh for Γ

fespace WhL(ThL,P1);
fespace PhL(ThL,P0);
fespace WhC(ThC,P1);
fespace Wh(Th,[P2,P2,P1],periodic=[[2,y],[4,y]]);

WhL fm,phi1L,phi2L,cgm,gm;
WhL c2gm;
```

With the macros and functions introduced in previous sections, the cost function is:

```
func real J(real[int] &X)
{
    fm[] = X;
    Th = movemesh(Th0,[x,fm+y*(H-fm)/H]);

    cgm[] = L2regul(fm[]); // Code 4.3

    stokes(); // Code 4.2
    return = -int1d(Th,3)(u1);
}
```

The derivative of the reduced cost function involves both state and adjoint macros defined in Codes 4.2 and 4.5. Before using the gradient interpolation macro written in Code 4.8, we perform a regularization of the functions ϕ_1 and ϕ_2 (Code 4.7).

```
func real[int] dJ(real[int] &X)
{
    fm[] = X;

    Th = movemesh(Th0, [x,fm+y*(H-fm)/H]); // Ω0 ↦ Ω
    ThC = movemesh(ThL, [x,fm+y*(H-fm)/H]); // (0,1) ↦ Γ

    cgm[] = L2regul(fm[]); // Code 4.3

    stokes(); // Code 4.2
    adjoint(); // Code 4.5

    phi1L[] = L2phiregul(1); // Code 4.7
    phi2L[] = L2phiregul(2); // Code 4.7

    gradInterp(); // Code 4.8
    return dJ;
}
```

The constraint introduced in Sect. 4.2 and its Jacobian are computed with the matrices introduced in Code 4.6:

```
func real[int] C(real[int] &X)
{
    real[int] cont(1+PhL.ndof+WhL.ndof);
    cont[0] = Cvol'*X; // ∫_0^1 f(x) dx = 0
    cont(1:PhL.ndof) = mDx*X; // |f'(x)| ⩽ M1
    cont(PhL.ndof+1:PhL.ndof+WhL.ndof) = mDxx*X; // |f''(x)| ⩽ M2
    return cont;
}

matrix dc;
func matrix jacC(real[int] &X)
{
    real[int,int] dcc(1,WhL.ndof); dcc = 0.0;
    dcc(0,:) = Cvol;
    dc = dcc;
    dc = [[dc],[mDx]];
    dc = [[dc],[mDxx]]; // [volume, |f'| ⩽ M1, |f''| ⩽ M2 ]
    return dc;
}
```

In FreeFEM, the possibility of using matrices and arrays instead of solving the variational formulation usually guarantees a more efficient algorithm in terms of execution speed and required memory. We wrote most of macros and functions with the explicit formulation for the sake of clarity, but the execution is quicker when dealing with matrices. Finally, it remains to call the optimization routine IpOpt:

```
real[int] start(WhL.ndof);
real[int] xub(WhL.ndof);
real[int] xlb(WhL.ndof);
real[int] cub(1+PhL.ndof+WhL.ndof);
real[int] clb(1+PhL.ndof+WhL.ndof);

// Variables bounds
xub=  XUB;
xlb= -XUB;
cub(1:PhL.ndof)=  CUB;
clb(1:PhL.ndof)= -CUB;
cub(PhL.ndof+1:PhL.ndof+WhL.ndof)=  HUB;
clb(PhL.ndof+1:PhL.ndof+WhL.ndof)= -HUB;

xub[0]  = 0.0; // f(0)=0
xlb[0]  = 0.0; // f(0)=0
xub[WhL.ndof-1]  = 0; // f(1)=0
xlb[WhL.ndof-1]  = 0; // f(1)=0
clb[0]  = 0.0; // ∫_0^1 f(x) dx = 0
cub[0]  = 0.0; // ∫_0^1 f(x) dx = 0

clb[PhL.ndof+WhL.ndof]  = -HUB/alpha;
cub[PhL.ndof+WhL.ndof]  = HUB/alpha;
clb[PhL.ndof+1]  = -HUB/alpha;
cub[PhL.ndof+1]  = HUB/alpha;

// intialization
WhL X0=0.0125/2*sin(x*pi*2*2);
if (hotrestart){
start = HOTRESTART(hotrestart); }

IPOPT(J,dJ,C,jacC,start,lb=xlb,ub=xub,clb=clb,cub=cub,tol=1.e-8);
```

Code 4.10 Calling IpOpt in the Stokes problem (see **shape_microswimmer.edp**)

The optimal solution returned by IpOpt is plotted in Fig. 4.4 for a rough and for a fine mesh. As mentioned above, the constraints on the first and second derivatives of f have been introduced to guarantee well-posedness and a good numerical convergence of the optimization process. As expected, the optimal solution saturates these constraints.

Since PDE optimization problems involve a large number of variables, iterations of the algorithm require a larger computational time as the mesh is finer (Fig. 4.5).

Having a good initial guess is also important. *Hot-restart* loops turn out to be effective by providing a better initialization that we refine by first running the algorithm on a rougher mesh and then on a finer one. An interpolation on the finer mesh of the solution obtained on the rough mesh is taken as a new initialization, so that the algorithm converges in a better way on the fine mesh. The hot-restart procedure is described in Appendix A.1. The whole code is available at FreeFEM's website.

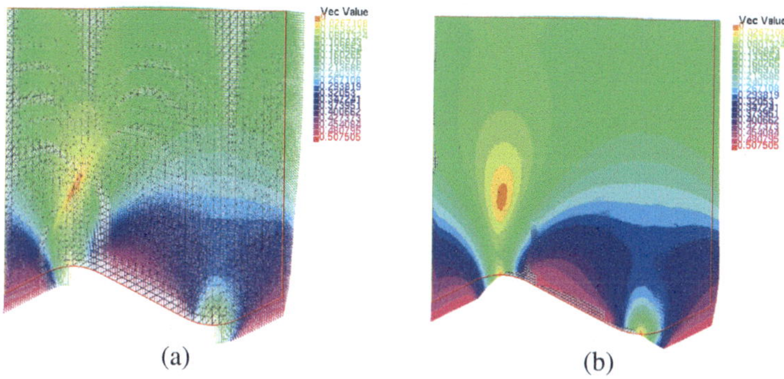

Fig. 4.4 Optimal solution of (4.1, 4.2, 4.3, 4.4, 4.6) for $M_1 = 0.4$ and $M_2 = 5.0$ on: (**a**) rough mesh; (**b**) fine mesh

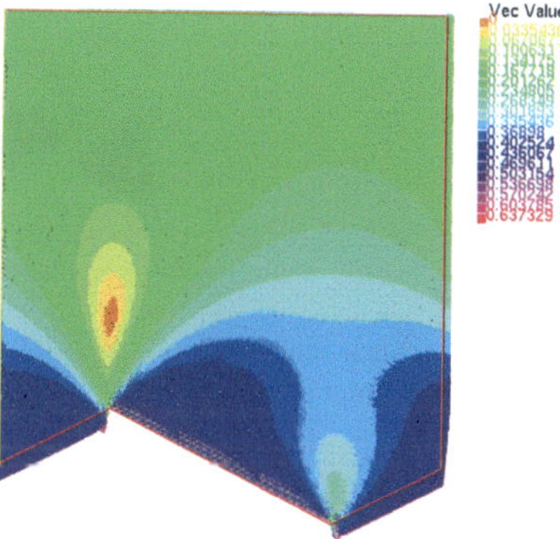

Fig. 4.5 Solution with no curvature constraint ($M_2 = +\infty$) and $J \approx 0.196$

4.5 Further Comments

We plot in Fig. 4.6 the optimal solutions for various values of M_2. We observe that the value of the cost functional at the optimal solution increases when M_2 is taken larger and that, as $M_2 \to +\infty$, the sequence of optimal solutions seems to converge to a triangular-shaped function, which may be the optimal solution of the problem when, formally, $M_2 = +\infty$, i.e., f varies in $H^1(0, 1)$ instead of $H^2(0, 1)$ without any constraint on f''. However, although this limit problem seems to be tractable from the numerical point of view, treating it rigorously from the theoretical point of view is much more difficult because existence of solutions of the Stokes problem (4.1, 4.2, 4.3, 4.4) is not guaranteed in $H^1(\Omega) \cap L^2(\Omega)$: it is required to consider other functional spaces (see [3, 8]) and the framework becomes much more complicated. We leave this issue as an interesting open problem.

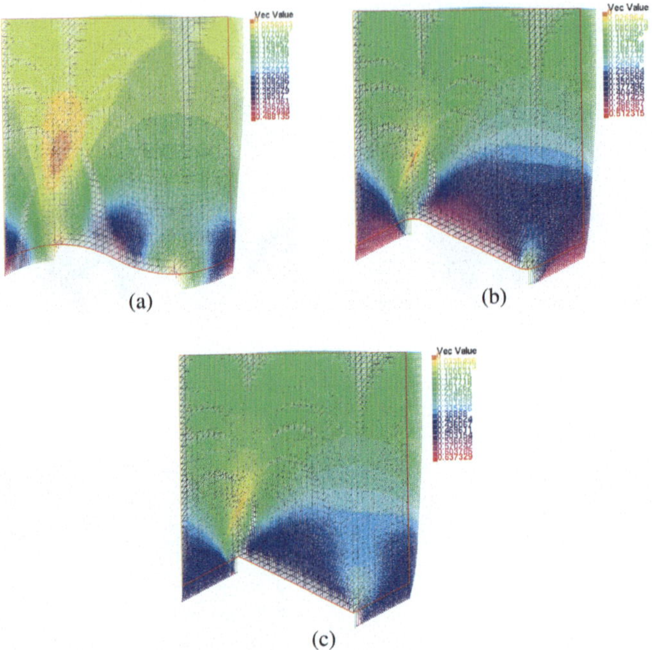

Fig. 4.6 Optimal solution of (4.1, 4.2, 4.3, 4.4, 4.6) for $M_1 = 0.4$ and: (**a**) $M_2 = 2$ and $J \approx 0.07$; (**b**) $M_2 = 10.0$ and $J \approx 0.16$; (**c**) $M_2 = 50.0$ and $J \approx 0.19$

Numerically, we can proceed as follows. Ignoring the second derivatives of f involved in the constraint function `C` and in Codes 4.8 and 4.9, the control f now varies in the subset of $U = H^1(0, 1)$

$$\mathcal{U}_{ad} = \left\{ f \in H_0^1(0, 1), \quad \int_0^1 f(x)\,dx = 0 \text{ and } |f'(x)| \leqslant M_1, \text{ for a.e. } x \in (0, 1) \right\}.$$

The optimal solution returned by `IpOpt` is plotted in Fig. 4.5.

References

1. Grégoire Allaire. 2007. *Conception optimale de structures*, volume 58 of Mathématiques & applications (Berlin) [Mathematics & applications]. Berlin: Springer. With the collaboration of Marc Schoenauer (INRIA) in the writing of Chapter 8.
2. François Alouges, Antonio Desimone, and Aline Lefebvre-Lepot. 2008. Optimal strokes for low Reynolds number swimmers: An example. *Journal of Nonlinear Science* 18: 277–302.
3. Chérif Amrouche and Vivette Girault. 1991. On the existence and regularity of the solution of Stokes problem in arbitrary dimension. *Proceedings of the Japan Academy, Series A: Mathematical Sciences* 67(5): 171–175.

4. Franck Boyer and Pierre Fabrie. 2013. *Mathematical tools for the study of the incompressible Navier-Stokes equations and related models*, volume 183 of Applied mathematical sciences. New York: Springer.
5. Frédéric De Gournay. 2006. Velocity extension for the level-set method and multiple eigenvalues in shape optimization. *SIAM Journal on Control and Optimization* 45(1): 343–367.
6. Vivette Girault and Pierre-Arnaud Raviart. 1986. *Finite element methods for Navier-Stokes equations*, volume 5 of Springer series in computational mathematics. Berlin: Springer. Theory and algorithms.
7. Antoine Henrot and Michel Pierre. 2018. *Shape variation and optimization*, volume 28 of EMS tracts in mathematics. Zürich: European Mathematical Society (EMS).
8. Marius Mitrea and Matthew Wright. 2012. Boundary value problems for the Stokes system in arbitrary Lipschitz domains. *Astérisque* (344): viii+241.
9. Bijan Mohammadi and Olivier Pironneau. 2001. *Applied shape optimization for fluids*. Numerical mathematics and scientific computation. New York: The Clarendon Press/Oxford University Press/Oxford Science Publications.
10. Shawn Walker and Eric Keaveny. 2013. Analysis of shape optimization for magnetic micro-swimmers. *SIAM Journal on Control and Optimization* 51: 3093–3126.

Appendix A
Supplements

A.1 Some FreeFEM Functions

The macro to build the matrix of jumps of the derivatives of the finite element basis functions is:

```
macro MatJumpofDx(Vh,Th,A)
{
    if (A.n) A.clear;
    matrix Adx(Vh.ndof,Th.nt);
    fespace Ph(Th,P0);
    matrix Dx = interpolate(Ph,Vh,op=1);
    assert(Vh.ndofK==2);
    int nt = Th.nt;
    for(int k=0; k< nt;++k)
    {
     Adx(Vh(k,0),k)=+1;
     Adx(Vh(k,1),k)=-1;
    }
    A = Adx*Dx;
} //
```

Code A.1 Matrix of derivative's jumps (see **shape_microswimmer.edp**)

Hot-restart requires first to use the option `warm_start_init_point yes` in the `IpOpt` option file *"optfile.opt"*. It is assumed that the code has already

F. Hecht et al., *PDE-constrained optimization within FreeFEM*, SpringerBriefs on PDEs and Data Science, https://doi.org/10.1007/978-981-95-7049-2

been run so that the files "Th0old.msh" and " fsol . txt " already exist. Then, it suffices to add the following lines at end of the code:

```
if (hotrestart == 0){
    savemesh(Th0,"Th0old.msh");
    {
    ofstream file("fsol.txt");
    file << X0[];
    }
}
```

The interpolation of a solution or of an initialization function only requires to provide a new mesh `ThL` and a finite element space `WhL` with the following routine:

```
func real[int] HOTRESTART(bool &hot)
{
    if (hot){
        WhL Xinit;
        mesh Th00 = readmesh("Th0old.msh"); // Initial mesh
        meshL ThL0 = extract(Th00,refedge=11);
        ThL0 = change(ThL0, flabel = fclab); // Straight Mesh for
            hotrestart
        fespace WhL0(ThL0,P1);
        WhL0 X00;
        {
        ifstream file("fsol.txt");
        file >> X00[];
        }
        Xinit = X00; // interpolation on the new mesh
        return Xinit[];
    }
}
```

Code A.2 Hot restart routine (see **shape_microswimmer.edp**)

Hot-restart is easy to implement in `FreeFEM`, thanks to the facility of interpolating from one given mesh to another one by just typing `ustart=uh1` with `ustart` and `uh1` respectively defined on the meshes `Th2` and `Th1`. An example can be found in **lq_stationary_O2_hotrestart.edp**. The user stipulates the int `hotrestart=0` to build the first solution and `hotrestart=1` to launch the algorithm on a finer mesh initializing with the previous solution.

The $L^2(\Omega)$ norm for functions defined on a finite element space V_h is coded as follows:

```
func l2norm(real[int] &X)
{
    // Th,Vh already constructed
    Vh u,v;
    varf varl2norm(u,v) = int2d(Th)(u*v);
    matrix M = varl2norm(Vh,Vh); // better to be constructed
        outside
    real[int] uu = M*u[];
    l2error = sqrt(X'*uu); // = (∫_Ω u(x)^2 dx)^(1/2)
    return l2error;
}
```

A.2 Semi-automatic Differentiation and Adjoint Method

In this section, we show that the adjoint code of the reduced cost functional of the problem (1.31, 1.30, 1.32) implicitly makes appear a discretization of the adjoint equation obtained by applying the Pontryagin maximum principle (see [9, Chapter 7]):

$$\begin{pmatrix}\dot{x}\\ \dot{y}\end{pmatrix} = \begin{pmatrix}1 & 1\\ 1 & -1\end{pmatrix} + \begin{pmatrix}u\\ v\end{pmatrix} \qquad \begin{pmatrix}x(0)\\ y(0)\end{pmatrix} = \begin{pmatrix}1\\ 1\end{pmatrix}, \tag{A.1}$$

$$\begin{pmatrix}\dot{p}\\ \dot{q}\end{pmatrix} = \begin{pmatrix}-1 & -1\\ -1 & 1\end{pmatrix} + \begin{pmatrix}x-1\\ 0\end{pmatrix} \qquad \begin{pmatrix}p(T)\\ q(T)\end{pmatrix} = \begin{pmatrix}0\\ 0\end{pmatrix}, \tag{A.2}$$

$$\int_0^T (p(u-\phi) + q(v-\psi))\,dt \leqslant 0 \qquad \forall(\phi,\psi) \in \mathcal{U}_{ad}. \tag{A.3}$$

The variational inequality (A.3) comes from the maximization condition of the Hamiltonian of the optimal control problem and makes appear the duality pairing as the L^2-inner product

$$\left(D\widehat{J}\left(\begin{smallmatrix}u\\ v\end{smallmatrix}\right), \left(\begin{smallmatrix}\phi\\ \psi\end{smallmatrix}\right)\right)_{L^2(0,T)} = -\int_0^T (p\phi + q\psi)\,dt$$

involving the derivative of the reduced cost functional, which thus yields its gradient

$$\nabla\widehat{J}\left(\begin{smallmatrix}u\\ v\end{smallmatrix}\right) = -\begin{pmatrix}p\\ q\end{pmatrix}.$$

The reduced cost functional $\widehat{J}$ obtained by solving (A.1) is computed thanks to an implicit Euler discretization (with n time steps) with the matrices

$$A = \begin{pmatrix} 1-\delta t & -\delta t \\ -\delta t & \delta t + 1 \end{pmatrix}$$

and

$$A^{-1} = \frac{1}{1-2\delta t^2} \begin{pmatrix} \delta t + 1 & 1 \\ 1 & 1-\delta t \end{pmatrix} = \mu \begin{pmatrix} a & 1 \\ 1 & b \end{pmatrix}$$

with $\delta t = \frac{1}{n-1}$, so that (A.1) and (A.2) are respectively discretized according to implicit and explicit schemes as

$$\forall k \in \{0 ... n-2\}, \quad \begin{pmatrix} x \\ y \end{pmatrix}^{k+1} = A^{-1} \begin{pmatrix} x \\ y \end{pmatrix}^{k} + \delta t A^{-1} \begin{pmatrix} u \\ v \end{pmatrix}^{k+1} \quad \begin{pmatrix} x \\ y \end{pmatrix}^{0} = \begin{pmatrix} 1 \\ 1 \end{pmatrix}, \tag{A.4}$$

$$\forall k \in \{0 ... n-2\}, \quad \begin{pmatrix} p \\ q \end{pmatrix}^{k+1} = A \begin{pmatrix} p \\ q \end{pmatrix}^{k} + \delta t \begin{pmatrix} x_k - 1 \\ 0 \end{pmatrix} \quad \begin{pmatrix} p \\ q \end{pmatrix}^{n-1} = \begin{pmatrix} 0 \\ 0 \end{pmatrix}. \tag{A.5}$$

The cost (1.30) is then computed by discretization according to a rectangle rule:

```
1  def J(u,v):
2       x[0]=0,y[0]=0,cost=0 # initialization
3       a = 1+dt, b = 1-dt
4       mu = 1/(1-2*dt**2)
5
6       for k in range(1,n-1): # dynamic loop
7           x[k] = mu*(a*x[k-1]+y[k-1]) + dt*mu*(a*u[k]+v[k])
8           y[k] = mu*(x[k-1]+b*y[k-1]) + dt*mu*(u[k]+b*v[k])
9           cost+= 0.5*(x[k]-1)**2
10
11      return cost
```

Following the successive steps for adjoint code generation presented in section "Automatic Differentiation", we write the derivative of the previous function obtained by automatic differentiation in reverse mode. To do so, some adjoint variables l_x, l_y, l_u, l_v, l_c are introduced, respectively related to the variables $x, y, u, v, cost$. The key is to range from the final step back to the first steps and to derive the computation line with respect to the variable involved and to finally update the corresponding adjoint variable. For example, the line 8

`(y[n-k]=mu*(x[n-k-1]+b*y[n-k-1])+dt*mu*(u[n-k]+b*v[n-k]))`

will generate lines 23 to 26 for updating the variables

```
(lx[n-k-1],ly[n-k-1],lu[n-k-1],lv[n-k-1])
```

```
1  def gradJ(u,v):
2      x[0]=0,y[0]=0,cost=0 # initialization
3      a = 1+dt, b = 1-dt
4      mu = 1/(1-2*dt**2)
5
6      for k in range(1,n-1): # dynamic loop
7          x[k] = mu*(a*x[k-1]+y[k-1]) + dt*mu*(a*u[k]+v[k])
8          y[k] = mu*(x[k-1]+b*y[k-1]) + dt*mu*(u[k]+b*v[k])
9          cost+= 0.5*(x[k]-1)**2
10
11     lx = 0, ly = 0, lc = 0 # arrays of size n
12     lc[n-1] = 1 # reverse mode initialization
13
14     for k in range(1,n-1):
15         lc[n-k-1] += lc[n-k]            # line 9: cost[n-k]
16         lx[n-k] += lc[n-k]*(x[n-k] - 1)
17
18         lx[n-k-1] += mu*ly[n-k]        # line 8: y[n-k]
19         ly[n-k-1] += b*mu*ly[n-k]
20         lu[n-k] += dt*mu*ly[n-k]
21         lv[n-k] += dt*b*mu*ly[n-k]
22
23         lx[n-k-1] += a*mu*lx[n-k]      # line 7: x[n-k]
24         ly[n-k-1] += mu*lx[n-k]
25         lu[n-k] += dt*a*mu*lx[n-k]
26         lv[n-k] += dt*mu*lx[n-k]
27
28     return lu,lv # Gradient
```

Having updated the adjoint variables, we have, for every $k \in \{0..n-2\}$,

$$\begin{pmatrix} l_x \\ l_y \end{pmatrix}^k = A^{-1} \begin{pmatrix} l_x \\ l_y \end{pmatrix}^{k+1} + \begin{pmatrix} x_k - 1 \\ 0 \end{pmatrix}, \qquad \begin{pmatrix} l_u \\ l_v \end{pmatrix}^k = \delta t A^{-1} \begin{pmatrix} l_x \\ l_y \end{pmatrix}^k$$

which gives, considering the variables l_u, l_v,

$$\begin{pmatrix} l_u \\ l_v \end{pmatrix}^k = A^{-1} \begin{pmatrix} l_u \\ l_v \end{pmatrix}^{k+1} + \delta t A^{-1} \begin{pmatrix} x_k - 1 \\ 0 \end{pmatrix} \quad \begin{pmatrix} l_u \\ l_v \end{pmatrix}^{n-1} = \begin{pmatrix} 0 \\ 0 \end{pmatrix}. \tag{A.6}$$

Since $\binom{p}{q}^k = -\binom{l_u}{l_v}^k$, we recognize an implicit Euler time discretization of (A.2), like for the state equation (A.1) in the backward direction which is to be compared with the explicit discretization (A.5) in forward direction. The derivative of the reduced cost function is therefore expressed as

$$-(p_0, ..., p_{n-1}, q_0, ..., q_{n-1})$$

according to the chosen discretization of the integral in (1.30, A.3).

A Crank-Nicolson scheme could have been chosen in (A.1). This would give a matrix A such that $A^{-1} = A$ and a Crank-Nicolson scheme for (A.2) too. We refer to [6] for a discussion on the relationship of automatic differentiation in reverse mode with derivative computation using adjoint variables, and resulting appropriate choices of discretization schemes.

A.3 PDE Optimization with `Python` or `Matlab`

Section "Automatic Differentiation" was devoted to automatic differentiation within `AMPL` for PDE constrained problems. Possible alternatives are either to write our own adjoint in `C++` code by calling routines like `CppAD` and `Adept` or to export the problem in `Python` or `Matlab` by using the `CasADi` package. Loading data into `Python` is done by:

```
ndof = int(numpy.loadtxt('ndof.txt'))
iA, jA, AM = numpy.loadtxt('A.txt', unpack=1, skiprows=3);
iM, jM, MM = numpy.loadtxt('B.txt', unpack=1, skiprows=3);
ydArray = numpy.loadtxt('target.txt',unpack=1,skiprows=1);

Ac = scipy.sparse.coo_matrix((AM, (iA, jA)), shape=(ndof,ndof));
Mc = scipy.sparse.coo_matrix((MM, (iM, jM)), shape=(ndof,ndof));
```

Code A.3 Sparse matrices importation in Python

The three first lines stand for the characteristics of the matrices and are skipped by using the option `skiprows=3` in the `numpy.loadtxt` command. The optimization package `CasADi` (see [1]) is then called and the problem is written as a nonlinear programming problem via the toolbox

```
import casadi as ca
opti = ca.Opti()
```

Declaring the variables is done by using the MX symbolics (assuming that most of operations between multiple sparse-matrix valued objects are allowed) whose description is available on the `CasADi` website https://web.casadi.org/docs/#the-mx-symbolics.

```
A = ca.MX(Ac); # convert sparse matrices above in MX object
M = ca.MX(Mc);
yd = ca.MX(ydArray) # target y_d in (2.1)

y = opti.variable(ndof) #Variables declaration
u = opti.variable(ndof)
```

It remains to express the cost and constraint functions. Matrix multiplication Ax is done either by `mtimes(A,x)` or `A @ x` commands. Most of possible operations are described in [2, Docs tab].

```
opti.minimize( 0.5*(y.T-yd.T) @ M @ (y-yd) + 0.5*alpha*u.T @ M @
    u ) #
    (2.1)
opti.subject_to( A @ y - M @ u == 0.0 ) # PDE constraint function
    (2.7)
opti.subject_to( opti.bounded(0,Mu,1) ) # Bounds on control
    0 ⩽ u ⩽ 1

opti.solver('ipopt',{'ipopt':{'max_iter':50, 'tol':1.e-11,
           'Hessian_approximation':'limited-memory'}})

sol = opti.solve()
```

Code A.4 `CasADi` template for LQ PDE Optimization (see **lq_stationary_casadi.edp**)

Computing the Hessian by automatic differentiation is in general too much memory greedy and too long in the execution, and BFGS thus appears as an alternative by using the `IpOpt` option

`'Hessian_approximation':'limited-memory'`

The previous code is written in the setting of **Option 1**. But **Option 2** can also be considered:

```
u = opti.variable(ndof) # only control as variable
us = M @ u
y = solve(A,us) # Solve state equation

cost = 0.5*(y.T-yd.T) @ M @ (y-yd) + 0.5*alpha*u.T @ M @ u
F = Function('F',[u],[cost]) # way to declare u ↦ Ĵ(u)

opti.minimize( F(u) )
opti.subject_to( opti.bounded(0,Mu,1) )
```

Appendix B
Identification of Parameters with `FreeFEM`

So far we have proposed differentiable optimization strategies for PDE-constrained optimization. For some reasons, we sometimes have to choose an alternative: the involved functions may fail to be differentiable and in such a case we are not able to write and use a gradient, or the cost function may also be non-convex with a large number of local extrema. We therefore need a robust derivative-free method to skim through both of these difficulties.

Derivative-free optimization is a domain that gathers many algorithms. Among the best known, genetic algorithms are often used for shape optimization, bayesian methods are sometimes preferred for image processing, and now we also have the growth of machine learning algorithms with a large number of applications. All these methods may be good candidates. In this appendix, we focus on the stochastic optimization algorithm `CMA-ES` that is implemented in `FreeFEM` and we seek to identify the Lamé parameters of a given beam whose behavior is given by a large number of measurements acquired through several experiments. We thus look for the parameters of the theoretical model, which is subject to the resolution of an elasticity PDE, in order to characterize the observed behavior.

B.1 The System of Elasticity

We focus on a two-dimensional beam and we want to determine its Lamé parameters thanks to several measurements of the boundary deformation. The uncertainties of the measurements are such that we cannot rely on a single measurement of the deformed beam. Therefore, we will consider many measurements of the deformed beam and we try to determine the theoretical deformation of the beam, so that it is on average as close as possible to the available measurements. The procedure is as follows: the left edge of the beam is fixed at its end and bends downwards due to the gravity field. We then collect a given number of measurements of the bent beam and we compare them to the theoretical deformation of the beam by solving an elasticity

F. Hecht et al., *PDE-constrained optimization within FreeFEM*, SpringerBriefs on PDEs and Data Science, https://doi.org/10.1007/978-981-95-7049-2

PDE. Finally, we search for the best possible two parameters that minimize a sum of square functions standing for the distance of the theoretical coordinates of the boundary to the data.

More generally, we are interested in solid objects deformed under the action of certain applied forces. We model this phenomenon by an elasticity PDE. Indeed, a point in the solid, initially at a certain position (x, y) will arrive at a new position (X, Y) after a given time (we consider the PDE to be stationary by assuming that the beam has reached its equilibrium position). If we further assume that the displacements are small and that the solid is elastic, the displacement vector $u = (x - X, y - Y)$ satisfies the following PDE, better known as Hooke's law, which expresses a relation between the stress tensor:

$$\sigma_{ij}(u) = \lambda\delta_{ij}\nabla \cdot u + 2\mu\epsilon_{ij}(u)$$

and the strain tensor

$$\epsilon_{ij}(u) = \frac{1}{2}\left(\frac{\partial u_i}{\partial x_j} + \frac{\partial u_j}{\partial x_i}\right).$$

Here, λ and μ are the two Lamé's coefficients that we are looking for. They both describe the mechanical properties of the solid, and are more commonly related to the more physical constants E, the Young's modulus and ν, the Poisson's ratio

$$\mu = \frac{E}{2(1+\nu)}, \qquad \lambda = \frac{E\nu}{(1+\nu)(1-2\nu)}.$$

We thus look at beam, which is initially assimilated to a rectangular domain Ω with boundary $\partial\Omega = \cup_{i=1}^{4}\Gamma_i$ and which is subject to external forces f like the gravity field. Its displacement u satisfies the PDE

$$-\mu\Delta u - (\mu + \lambda)\nabla(\nabla \cdot u) = f.$$

Note that we do not use this equation because the associated variational form does not give the right boundary condition, we simply use:

$$\mathrm{div}(\sigma(u)) = f,$$

whose corresponding variational formulation reads

$$\int_\Omega \sigma(u) : \epsilon(v) - \int_\Omega f x \, dx = 0,$$

or

$$\int_\Omega \lambda\nabla u \cdot \nabla v + 2\mu\epsilon(u) : \epsilon(v) - \int_\Omega f x \, dx = 0. \tag{B.1}$$

We then introduce a mesh (see Fig. B.1) of the rectangular beam that we will modify through the vector displacement u obtained after having solved the previous variational form (B.1).

The experiment that we consider to collect measurements is the following: the left board Γ_4 is fixed while the gravity fields acts on the elastic beam such that it goes down. Boundary conditions are such that the left boundary is fixed thanks to a Dirichlet boundary condition

$$u = 0 \text{ on } \Gamma_4,$$

while the three other sides are free of constraints

$$\sigma(u) \cdot \mathbf{n} = 0 \text{ on } \Gamma_i, \; i \in \{1, 2, 3\}.$$

To numerically solve the elasticity PDE, we refer to [3]. We use $\mathbb{P}_2$ finite element. The solving procedure is written in Code B.1.

```
fespace Vh(Th, P2);
Vh u, v;
Vh uu, vv;

// Macro
real sqrt2=sqrt(2.);
macro epsilon(u1,u2) [dx(u1),dy(u2),(dy(u1)+dx(u2))/sqrt2] //
// The sqrt2 is because we want: epsilon(u1,u2)'* epsilon(v1,v2)
    - epsilon(u): epsilon(v)
macro div(u,v) ( dx(u)+dy(v) ) //

// Problem
real mu= E/(2*(1+nu));
real lambda = E*nu/((1+nu)*(1-2*nu));

macro pdelamemacro(){
    solve lame([u, v], [uu, vv])
        = int2d(Th)(
            lambda * div(u, v) * div(uu, vv)
          + 2.*mu * ( epsilon(u,v)' * epsilon(uu,vv) )
        )
        - int2d(Th)(
            f*vv
        )
        + on(4, u=0, v=0)
        ;
    } //
```

Code B.1 Elasticity solving with FEM (see **beam_paramater_identification.edp**)

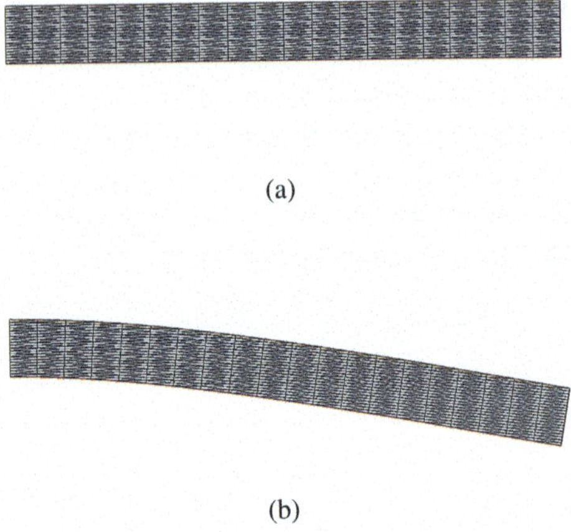

Fig. B.1 Mesh of the beam: (**a**) initial mesh; (**b**) and modified one

Code B.1 returns the displacement vector of the beam. To finally obtain the deformed beam, we need to move the initial rectangular mesh thanks to the vector displacement u. This step is performed by the command movemesh:

```
mesh Thm = movemesh(Th, [x+u, y+v]);
```

We then plot on Fig. B.1 the initial and the modified meshes. The modified beam is moved with a multiplicative coefficient in the movemesh command in order to well observe the deformation. Once the beam has been moved, we collect the coordinates of its boundary thanks to the following code:

```
int[int] ll = [1,2,3];
meshL Thl = extract(Th,refedge=ll); // extract boundary
real[int] coord(2*ThL.nv); // coordinates of the boundary

for (int i=0;i<ThL.nv;i++)
{
    coord[i] = ThL(i).x;
    coord[ThL.nv+i] = ThL(i).y;
}
```

For the sake of simplicity, we focus here on a linear elasticity model. Nevertheless, a nonlinear elasticity model, much more realistic, would be feasible at the cost of additional iterations necessary to solve the resulting variational formulation by a fixed point or Newton method as already described in Sect. 2.3.

B.2 Identification Problem

We now denote by $(X_{mes})_{i\in\{1..N\}}$ several measurements of the coordinates of the deformed beam boundary. We look for the two corresponding Lamé parameters that will give the closest X coordinates of the theoretical beam boundary, by minimizing a sum of quadratic functions. For this purpose, we consider an objective function that takes the Lamé parameters as unknowns, computes the coordinates of the beam modified by the elasticity PDE and then compares these coordinates to the available measurements. We thus consider the following cost function which computes the relative errors of the coordinates to the measurements

$$\min_{\lambda,\mu} \frac{1}{N} \sum_{i=1}^{N} \frac{\|X(\lambda,\mu) - X^i_{mes}\|^2}{\|X^i_{mes}\|^2} \tag{B.2}$$

such that $X(\lambda, \mu)$ includes the coordinates of the moved beam by the displacement vector u solution of:

$$\begin{cases} \operatorname{div}(\sigma(u)) = f, & \\ u = 0 & \text{on } \Gamma_4, \\ \sigma(u)\cdot \mathbf{n} = 0 & \text{on } \Gamma_i,\ i \in \{1,2,3\}, \end{cases} \tag{B.3}$$

where f includes the gravity vector field. We sum up the procedure in Fig. B.2 This one requires first to solve the elasticity PDE and then to move the mesh. This is summarized in the following Code B.2:

```
func real[int] beamcoord(real &lambda, real &mu)
{
    pdelamemacro();

    meshL ThL = movemesh(Th1,[x+eps*u,y+eps*v]); // move the
        boundary with [u,v] displacement

    real[int] coord(2*ThL.nv); // Coordinates of the boundary's
        vertices
    for (int i=0;i<ThL.nv;i++)
    {
        coord[i] = ThL(i).x;
        coord[ThL.nv+i] = ThL(i).y;
    }

    return coord;
}
```

Code B.2 Function returning the coordinates of the theoretical moved beam

We now implement a function that will measure, for given parameters λ and μ, the closeness of the theoretical beam boundary with the different measurements

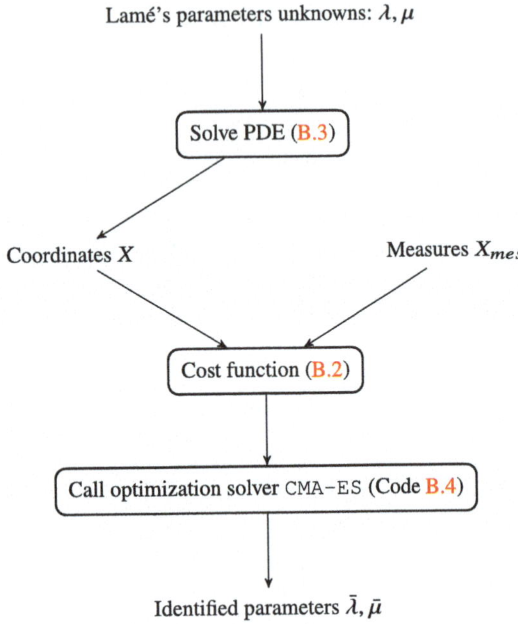

Fig. B.2 Recapitulative procedure of the identification problem

of the deformed beam boundary that we have. To do so, we quantify it by a least squares method, which is thus constrained by the resolution of the theoretical PDE (B.1). This function is written in Code B.3.

```
func real leastsquaremeasure(real[int] &X)
{
    real[int] coordX = beamcoord(X[0],X[1]);
    real objective = 0.;
    real objective0 = 0.;
    for (int j=0;j<target.n;j++)
    {
        objective0 = 0.;
        for (int i=0;i<n;i++)
        {
            objective0 += (noisytarget(j,i)-coordX[j])^2;
        }
        objective0 *= (1./(norm2(noisytarget(j,:))));
        objective += objective0;
    }
    objective *= (1./n);
    cout << "cout␣=␣" << objective << "␣||␣lambda␣=␣" << X[0] <<
        "||␣mu␣=␣" << X[1] << endl;
    return objective ;
}
```

Code B.3 Least square objective's function (see **beam_paramater_identification.edp**)

The resulting objective function is a sum of square functions that has many local minima. A gradient method or a differentiable method is not feasible because there is a high probabily that the iterates will get trapped around a local minimum and will not reach the global minimum. Hence, we are going to use a more robust algorithm which has the advantage of searching for a solution in a wider spectrum at the cost of a larger number of iterations and thus a slower convergence. We use the algorithm CMA-ES, a stochastic algorithm that only requires the cost function (not its derivatives). Its behavior and functioning are detailed in [5]. More information can also be found on FreeFEM's website https://freefem.org and the whole numerical code is available at FreeFEM's page.

```
real[int] Xinit = [50000,50000]; //initialization point
real[int] u0 = Xinit;
real[int] StDevs=[100000,100000]; //initial deviations
real min = cmaes(leastsquaremeasure,u0,stopTolFun=1e-5,
    stopMaxIter=1000,initialStdDevs=StDevs,seed=random(),popsize
    =25);
```

Code B.4 CMA-ES solver

Let us give some explanations on the options given by CMA-ES:

- seed: seed for random number generator.
- initialStdDevs: value σ for the standard deviations of the initial covariance matrix. If the value is passed, the initial covariance matrix will be set to σ Id. The expected initial distance between initial between unknowns and optimal solution should be roughly initialStdDevs. Here, we have two unknowns, thus we give a two dimensional array.
- popsize: integer value used to change the sample size. Increasing the population size usually improves the global search capabilities at the cost of, at most, a linear reduction of the convergence speed with respect to popsize.

B.3 Results

In our example, we are looking for the Lamé parameters of a beam whose Young's modulus and Poisson's ratio are $E_0 = 2100000$, $\nu_0 = 0.28$. This gives $\lambda_0 = 1044030$, $\mu_0 = 820312$. We have at our disposal several measurements of the beam. We can well imagine that pads have been placed on the boundaries (except the fixed boundary) of the beam at rest (these sensors coincide with the vertices of the mesh) and we measure their coordinates after deformation. This gives a number of data for which we then have several measurements in order to have on average a good approximation. At the end, the measurements are stored in a two dimensional array (here we have 62 measurements corresponding to the coordinates with 100 samples of each). We then use the algorithm described in Fig. B.2 several times (10 times) and we obtain on average the following Lamé coefficients $\bar{\lambda} = 1020260$, $\bar{\mu} = 825761$, with a mean relative error $\epsilon_\lambda = 2.3\%$, $\epsilon_\mu = 0.7\%$.

We thus almost find the initial Lamé's parameters with a relatively good accuracy considering the measurement uncertainties. CMA-ES seems to be a good alternative for optimization problems with some difficulties such as the computation of derivatives or a large number of local minima. We can also appreciate its ability to obtain a good estimate of a solution in order to initialize a more accurate and faster optimization method.

A good choice may be to perform the first iterations of the optimization with the robust algorithm CMA-ES, before switching, for example, to IpOpt that is very efficient to find local minima.

For readers wanting to go further, for instance Lamé's parameters emulation with artificial intelligence tools, we bring to the foreground the possibilities to make a link with FreeFEM and Python in Appendix A.3. Other possibilities are being developed, like the Python package "pyfreefem" [4]. Some tools of Deep Learning implemented in Python (tensorflow and pytorch modules) are thus available. Let us mention [8], where the authors use Deep Learning methods for parameter identification. For readers who want to find further details on PDE-optimization under uncertainties, we advise to skim through the book [7]. The authors propose a smooth transition from optimal control of deterministic PDEs to optimal control of random PDEs. It covers uncertainty modelling in control problems, variational formulation of random PDEs, existence theory and numerical resolution methods.

References

1. Joel A.E. Andersson, Joris Gillis, Greg Horn, James B. Rawlings, and Moritz Diehl. 2009. CasADi—a software framework for nonlinear optimization and optimal control. *Mathematical Programming Computation* 11(1): 1–36.
2. CasADi. https://web.casadi.org/docs/.
3. Alexandre Ern and Jean-Luc Guermond. 2021. *Finite elements II*. Berlin: Springer.
4. Florian Feppon. 2019. *Shape and topology optimization of multiphysics systems*. PhD thesis, Thèse de doctorat de l'Université Paris-Saclay préparée à l'Ecole polytechnique.
5. Nikolaus Hansen. 2016. The CMA evolution strategy: A tutorial. CoRR, abs/1604.00772.
6. Jesús María Sanz-Serna. 2016. Symplectic Runge-Kutta schemes for adjoint equations, automatic differentiation, optimal control, and more. *SIAM Review* 58(1): 3–33.
7. Jesús Martínez-Frutos and Francisco Periago Esparza. 2018. *Optimal control of PDEs under uncertainty. An introduction with application to optimal shape design of structures*. SpringerBriefs in Mathematics. Cham: Springer; Bilbao: BCAM – Basque Center for Applied Mathematics.
8. Olivier Pironneau. 2019. Parameter identification of a fluid-structure system by deep-learning with an Eulerian formulation. *Methods and Applications of Analysis* 26(3): 281–290.
9. Emmanuel Trélat. 2005. *Contrôle optimal*. Mathématiques Concrètes. Paris: Vuibert. Théorie & applications.

The manufacturer's authorised representative in the EU is Springer Nature Customer Service Centre GmbH, Europaplatz 3, 69115 Heidelberg, Germany. If you have any concerns regarding our products, please contact ProductSafety@springernature.com

Printed and bound by CPI Group (UK) Ltd, Croydon, CR0 4YY
07/07/2026
02160927-0006